EUGÈNE DUBOIS AND THE APE-MAN FROM JAVA

Eug Dubois

BERT THEUNISSEN
Institute for the History of Science
University of Utrecht, The Netherlands

Eugène Dubois and the Ape-Man from Java

The History of the First 'Missing Link' and Its Discoverer

Kluwer Academic Publishers
DORDRECHT / BOSTON / LONDON

Library of Congress Cataloging in Publication Data

Theunissen, Bert, 1955-
Eugène Dubois and the ape-man from Java : the history of the first missing link and its discoverer / Bert Theunissen.
p. cm.
Originally presented as the author's thesis (Ph. D.--1985) under the title: Eugène Dubois en de aapmens van Java.
Bibliography: p.
Includes index.

ISBN-13: 978-94-010-7491-9 e-ISBN-13: 978-94-009-2209-9

DOI: 10.1007/978-94-009-2209-9

1. Pithecanthropus erectus. 2. Dubois, Eugène, 1858-1940.
3. Anthropology, Prehistoric. 4. Fossil man--Indonesia--Java.
I. Title.
GN284.6.T48 1988
573.3--dc19 86-19223
CIP

First published as *Eugène Dubois en de aapmens van Java* by Rodopi, Amsterdam, 1985.
Translated from Dutch by Enid Perlin-West.

Published by Kluwer Academic Publishers,
P.O. Box 17, 3300 AA Dordrecht, The Netherlands.

Kluwer Academic Publishers incorporates
the publishing programmes of
D. Reidel, Martinus Nijhoff, Dr W. Junk and MTP Press.

Sold and distributed in the U.S.A. and Canada
by Kluwer Academic Publishers Group,
P.O. Box 332, 3300 AH Dordrecht, The Netherlands.

For Inge Nicolette

CONTENTS

CHAPTER 5: CEPHALISATION, PITHECANTHROPUS, AND EVOLUTION

SOURCES

FOREWORD

Although the name *Pithecanthropus* is now seldom used, there are few who study the origin of our species who will fail to recognise the historical place of the usage and its association with Eugène Dubois. During the last thirty or forty years, *Australopithecus* and its African context has tended to draw attention from the early work on our origins in Java. It is now increasingly common to hear the term 'pithecanthropine' used only to indicate the Asian or Far Eastern examples of *Homo erectus* which, although probably derived from African ancestry, have some features that in the opinion of some experts may justify their being considered distinctive. This discussion is not within the pages that follow which deal extensively with the work of Eugène Dubois. He was an extraordinary man who did as much as any person since to put the great antiquity of our ancestors firmly in the public domain.

Dubois became involved with the study of human origins from a medical and anatomical background as have many since. The jealousies and professional pressures that we think of as a phenomenon of the post-war years were clearly a major factor in deciding the future of his career. I found it fascinating to read of incidents which happened almost one hundred years ago that, except for the personalities, could well be a description of events taking place presently! Similarly, the professional and lay responses to some of his early work remind me of responses today following the announcement and publication of new discoveries and ideas. I feel better for knowing that one of the great pioneers in the field experienced the same difficulties that persist to the present but by the same token, I am saddened that so little has changed in spite of the great wealth of data that is now before us.

This book is a valuable contribution. It brings out a wealth of background and detail that has never before been presented in English and it will be a useful source for many scholars seeking to undertake the early work that was done in the Far East on human origins. The debate on Asia's role in the evolution of our species continues to be present and will probably be an open subject for some years yet. It is interesting to speculate on where Dubois would have stood in the presence of the well preserved early African *Homo erectus* material. The historical account of Eugène Dubois and the ape-man from Java makes such speculation entirely justified although I think his position would have been different depending upon the timing. This too might well be the same for many of us today but there is much to be learned from the fascinating account of the life and work of Eugène Dubois.

Richard Leakey

ACKNOWLEDGEMENT

I have benefitted from the help and advice of many people while writing this book and an earlier, Dutch version of it which served as my PhD dissertation. I wish to express my gratitude to all of them. Here I shall only mention those who have assisted me in producing this English version. Professor H.J.M. Bos, Professor C.A.W. Korenhof, Dr. G.N. van Vark and Dr. J. de Vos read and commented upon (parts of) the manuscript; I thank them for saving me from several blunders. Mrs. Enid Perlin–West translated the Dutch text; I greatly enjoyed our 'samenwerking'. Anneke Schoonhoven enabled me to see what I had written by typing and retyping the text. The help of Lian Hielkema in proofreading has been invaluable. Finally, the financial support, in the form of a translation grant, of the Dutch Organisation for Scientific Research (NWO) is gratefully acknowledged.

INTRODUCTION

To mention the name of Dubois is to say *Pithecanthropus erectus* in the same breath. Dubois gained international fame through his discovery of the 'upright-walking ape-man', and his investigation of this fossil hominid was the *pièce de résistance* of his scientific work. Everyone interested in human evolution and who has read anything on the subject can say, in broad outline, who Dubois was and what he did. The story of his discovery of Pithecanthropus has been told in greater or lesser detail in countless popular works and reference books on human descent, yet historians of science have devoted hardly any attention to Dubois. Delving into the existing literature one discovers all too soon that it fails to provide answers to many elementary historical questions, or that the answers are of a speculative and undocumented kind. There is no comprehensive, source-based analysis and interpretation of Dubois' palaeoanthropological research.

This study aims at filling this gap. Firstly, I shall provide an outline of Dubois' palaeoanthropological work and of his investigations in closely related areas. Secondly, I shall analyse the way in which Dubois' discovery and interpretation of the ape-man was received in scientific circles. Finally, the significance of Dubois' work for the development of palaeoanthropology will be examined.

The chapter division follows, in broad outline, the chronology of Dubois' life and work. I shall begin with a survey of the state of affairs in palaeoanthropology at the time Dubois entered the research field. This background is indispensable for a proper insight into Dubois' role and significance. Dubois himself will be brought on stage in chapter two. I shall describe his background and training, and try to reveal his motives for going in search of Man's fossil ancestors. Next we shall see how he gave shape to his plans. This chapter will show the extent to which Dubois' ideas fitted in with contemporary palaeoanthropological thought, and the extent to which he freed himself from tradition to plot a new course.

Dubois' discovery and first description of *Pithecanthropus erectus* will be discussed in chapter three. I shall devote particular attention here to the theoretical concepts underlying Dubois' view of his missing link between ape and human: how did he conceptualise his 'upright-walking ape-man' on the basis of such scanty fossil material? The fourth chapter deals with the way in which his account of the fossils was received in scientific circles. I shall analyse the discussions on Pithecanthropus and assess the historical significance of Dubois' find for the development of research into human descent.

Chapter five will deal with the development of Dubois' ideas on the ape-man over time. To this end it is necessary to tackle a subject that seems at first sight to have little to do with Dubois' palaeoanthropological work but which is, on closer inspection, inextricably linked with it. This is his research into cephalisation, that is, his study of the relationship between brain and body weights among vertebrate animals. The central thesis of this chapter is that Dubois' later view of Pithecanthropus derived from the results of his research on the brain. Further, there will be a discussion of Dubois' reaction to the new finds of *Pithecanthropus erectus* and *Sinanthropus pekinensis* made in the 1930s in Java and China. This will reveal that, towards the end of his life, his opinions placed him in an increasingly isolated position within palaeoanthropology.

From this statement of intent it will be clear that I do not aim at writing a biography; biographical information will only be given when this is necessary for a good understanding of Dubois' work. Unfortunately, there is still no full biography available.[1] A brief sketch of Dubois' life will permit the reader some grasp of biographical matters.

Marie Eugène François Thomas Dubois was born on 28 January 1858 at Eijsden in the province of Limburg, to the country apothecary Jean Joseph Balthasar Dubois (1832-1893) and Maria Catharina Floriberta Agnes Roebroeck (1830-1911). After passing through the Eijsden primary school he attended the State High School (*Rijks Hogere Burgerschool*) at Roermond. In 1877 he was registered at the University of Amsterdam as a student of medicine. In 1881 he became an assistant in the anatomical laboratory, and in the same year he was appointed as teacher of anatomy at the State Training School for Art Teachers (*Rijksnormaalschool voor Tekenonderwijzers*) and the State School of Applied Art (*Rijksschool van Kunstnijverheid*) in Amsterdam, filling both posts until 1887.[2] Dubois qualified as a medical doctor in 1884 and two years later, in 1886, he was appointed as lecturer in anatomy at Amsterdam University. In that year he married Anna Geertruida Lojenga (1862-1943) from Elburg.

In 1887 Dubois threw up his career in Amsterdam and enlisted for eight years as Medical Officer Second Class in the Royal Dutch East Indies Army (*Koninklijk Nederlandsch-Indisch Leger*) with the aim of mounting a search in the Dutch Indies for human ancestors. He was stationed on Sumatra from 1887 to 1890, then went to Java, where he discovered the remains of *Pithecanthropus erectus* in 1891 and 1892.

Returning from the East Indies in 1895 Dubois led a sedentary life in the Netherlands until his death. After living in The Hague for a short while he finally settled in Haarlem in 1897, in which year the University of Amsterdam awarded him an honorary doctorate in botany and zoology. Two years later the same university appointed him as professor extraordinary in crystallography, mineralogy, geology, and palaeontology. A full professorship in these subjects followed in 1907, physical geography being added at the same time to his teaching duties; he taught this subject until 1921. From 1897 until his death Dubois was also curator of Teylers Museum in Haarlem.

Besides his research into human descent and related subjects, Dubois investigated a wide variety of scientific problems, particularly in the first two decades of this century.[3] To mention the most important, he took up research on the Teglian Clay, a formation of fundamental importance for both geology and palaeontology, which marks the border between the Pliocene and Pleistocene. Under Dubois' guidance two of his students, A. Schreuder and J.J.A. Bernsen, wrote their doctoral theses on the fossil fauna of Tegelen.[4] Dubois also studied climate in the geological past, and in 1891 he posited the theory that climatic changes in the earth's history were connected with the evolution of the sun. He investigated the carbon cycle and produced an estimate of the earth's age on the basis of his findings. He delved into the genesis of the Dutch sand dunes and linked their formation with that of the Straits of Dover. He also published on the origin of the Dutch peat bogs and fens. Finally, he contributed to the discussion on the drinking-water supply in the coastal provinces with studies on ground water and dune water.[5]

In 1928 Dubois officially retired, but owing to lack of a successor he only relinquished his teaching duties in 1929. Scientifically he remained active until his death. For the last years of his life, from 1937 on, he lived on his estate 'De Bedelaar' near Haelen in central Limburg, which he had purchased in 1906 and had gradually transformed into what he called a 'nature park' over the years.[6] Dubois died there on 16 December 1940.

Notes

1 Useful sources are Barge, "In memoriam Eugène Dubois"; Boeke, "Levensbericht"; Brongersma, "De verzameling van Indische fossielen", "Eugène Dubois" and "Een geologisch verleden"; Polak, "Eugène Dubois"; Tesch and Brongersma, "Eugène Dubois".

2 From 1884 on Dubois was first assistant (prosector) in the anatomy laboratory.

3 There is a concise discussion in Polak, "Eugène Dubois".

4 Schreuder, *Conodontes*; Bernsen, *Geology of the Teglian Clay*. In fact Schreuder and Bernsen were his only graduates.

5 Dubois' most important contributions to hydrology are discussed in De Vries, *Hydrologisch onderzoek*, chapter 3, especially 42-46, 51-54, 58.

6 For this see Kruytzer, "Bezoek aan 'De Bedelaar'"; Lemaire, "De Bedelaar van prof. Dubois".

CHAPTER 1

THE NINETEENTH-CENTURY BACKGROUND

THE THEORY OF EVOLUTION AND THE POSITION OF MAN

In his *Origin of Species* Charles Darwin was very reticent on the question of human descent. He had already considered the problem of Man by the end of the 1830s, when the principle of evolution through variation and natural selection took shape in his mind,[1] but all we read about our species in the *Origin*, published in 1859, is that: "Light will be thrown on the origin of man and his history."[2] In spite of this caution the implications of evolutionary theory for Man were immediately clear to the attentive reader: humans belonged to the animal kingdom, and had developed from lower forms by a gradual process of variation and natural selection. Many people also came to the immediate conclusion that our far-distant ancestors must be sought within the line of primates.

The question of human ancestry was central to many discussions of Darwin's theory, and assaults on his ideas were frequently launched in resistance to the consequences of evolutionary theory for Man. Darwin was undermining the hoary old conviction that *Homo sapiens* differed fundamentally from the animals, in the intellectual sense if not in the physical. The attack on the idea of this superior status and all the social, ethical, and particularly religious norms and values linked with it, was unacceptable to many people.

The position of the human species also provided the main issue in the famous debate between Richard Owen and Thomas Henry Huxley, Owen thinking that humans had to be distinguished from the apes on the basis of biological characteristics. According to Owen several sections of the human brain, including the hippocampus minor, did not occur in the other primates, and this reflected the superiority of human intelligence. He found this sufficient argument for setting Man apart from the apes. Huxley successfully disputed this statement, showing that the brain sections Owen had cited were in fact present in apes.[3]

The debate was decided overwhelmingly in favour of Huxley, but objections to the application of evolutionary theory to Man were not thereby overcome. Darwin's thesis that the animal kingdom had developed by a process of evolution was accepted fairly quickly, but there was to be much further debate over the position of mankind. Even when biological counter-arguments appeared to have been silenced, there still remained the problem of the human mind, which for many people was a phenomenon separating humans from animals, and which could not be explained by presuming a process of evolution.

It is telling that even the co-author of the evolutionary theory, Alfred Russell Wallace, placed Man in a special position.[4] Wallace originally merely introduced a few restrictions concerning the scope of the natural selection mechanism. He asserted that the mind had governed human evolution after the human body had more or less acquired its final form. Man had not adapted to his environment through physical change caused by variation and natural selection, but by means of his mental capacities.[5] Later on, Wallace went even further: the human mind could not possibly have come into existence through natural selection, and he regarded this mechanism as only partly effective even where the body was concerned.[6] In later years Wallace inclined increasingly toward the view that humankind must have originated rather suddenly, guided by a superior, purposeful force or power.[7]

Even Charles Lyell, whose *Principles of Geology* had contributed so greatly to the formation of Darwin's thinking, and who had hesitantly endorsed the theory of evolution after his original resistance to it, retained ambivalent views on the question of human descent. In this Lyell's motives were to a certain extent religious ones.[8]

In his *Descent of Man*, published in 1871, Darwin introduced the principle of sexual selection as an additional mechanism to explain the process of evolution; in point of fact, the *Descent* was more a book on the functioning of sexual selection than on human origins. Yet where Darwin discussed the position of Man within the animal kingdom, he carried his line of thinking to its logical conclusion that humankind had developed by a gradual process from more ape-like ancestors. Human evolution, both physical and mental, could be explained by natural causes.

We have seen that others had already debated the implications of the *Origin* for Man. As early as the beginning of the 1860s, several militant supporters of Darwin published essays discussing the position of the human species from an evolutionary viewpoint, Huxley being one of the first in 1863. In his *Man's Place in Nature* he demonstrated, particularly on the basis of comparative anatomical and embryological data, that there were no essential anatomical differences between humans and the apes. Huxley was demolishing the barrier that, according to Owen, lay between Man and the rest of the animal world. The most important conclusion drawn from his research was that "the structural differences which separate Man from the Gorilla and the Chimpanzee are not so great as those which separate the Gorilla from the lower apes."[9] This was to become an often repeated dictum among adherents of Darwin and Huxley. *Homo sapiens* had to be counted as part of the animal kingdom on anatomical grounds, and consequently there could be no objection to applying evolutionary theory to humans. Although some, Theodor Bischoff and Christian Aeby for instance, contested Huxley's conclusions, these were confirmed time and again by morphological investigation, and people soon began to talk of 'Huxley's Law'.[10]

In Germany the monistic-materialist Ernst Haeckel emerged as Darwin's most influential champion, regarding evolution as a proven and indisputable fact. For him, Man's position was a purely zoological question, and he intended to reconstruct the human genealogical tree in broad outline with the aid of comparative anatomical and embryological data, despite the current lack of material palaeontological evidence. In his *Natürliche Schöpfungsgeschichte* of 1868 he represented this family tree as a monophyletic progression of twenty-two stages, beginning with the most primitive organisms and ending with modern humans. According to Haeckel, mankind had evolved from a line of primates during the Tertiary period, in the Pliocene or Miocene. The origins of Man were based upon acquisition of the ability to walk in an upright position and of the capacity for speech, and upon an increase in the size of the brain.

Haeckel interpolated an intermediary stage, the Pithecanthropi (ape-men) or Alali (speech-less) between the apes which formed the twentieth stage in the process of human evolution, and the modern human, the twenty-second stage. He argued that these beings already had the human habitus, but that they still lacked articulated speech, Man's most characteristic feature. He suggested *Pithecanthropus alalus*, the speechless ape-man, as hypothetical representative of this stage, and he even indicated where humankind had originated – Lemuria, a sunken continent that had once stretched from the coast of Africa over the Indian Ocean to the Philippines.[11] Haeckel was not lacking in imagination; even his contemporaries felt that he sometimes indulged too freely in speculation.

Carl Vogt was another materialist who tried to prove the anthropoid origin of humankind. In a controversial series of lectures in which he made no attempt to conceal his disdain for religious sensibilities, Vogt posited the thesis that the defects apparent in microcephaly (the possession of an abnormally small brain) offered proof that humans had descended from apes. He considered the microcephalic to be a kind of living missing link, arguing that the brains of people afflicted by this disorder were so retarded in their development that they represented, both quantitatively and qualitatively, a transitional form between the ape brain and the human.[12] The theory caused a sensation for a short time, but wilted rapidly under the criticism of Bischoff and Aeby, for instance.[13] It is nevertheless significant that Vogt, reaching a wide public with his lectures, located Man unconditionally within the animal kingdom, and did not exclude him from the process of evolution. We can find other examples of this position in the works of Ludwig Büchner and Friedrich Rolle, who also published their views well in advance of the appearance of Darwin's *Descent of Man*.[14]

It was the avowed materialists among biologists and anthropologists who developed the most explicit ideas of human evolution, while more positivistically inclined scientists adopted a more cautious attitude. They regarded the theory of evolution strictly as an hypothesis still to be tested; they preferred to set aside speculation about human origins and confine their attention to the empirical. This was a particularly marked attitude among anthropologists and, as we shall see, it was the anthropologists above all who were to determine the general appearance of palaeoanthropology in the second half of the nineteenth century.

In the 1860s there were lively debates within the Paris *Société d'Anthropologie* over the theory of evolution and the question of human descent.[15] Harvey distinguishes both positivist and materialist currents among French anthropologists,[16] with the positivists at first gaining the ascendence within the *Société*, the materialists acquiring greater influence only after 1880. The positivists laid stress on observation, description, and measurement, preferring an inductive working method; they felt that they should refrain from theoretical, and certainly from speculative considerations. For them Darwin's theory of evolution was an unproven hypothesis, and only empirical research would finally determine its value. In contrast the materialists accepted evolutionary theory as a given, and applied it unconditionally to Man. For example, in 1873 Gabriel de Mortillet and Abel Hovelacque drew up a human family tree in which they placed, between the apes and Man, a hypothetical intermediate form they named *l'Anthropopithèque*, to which they attributed characteristics somewhere between those of apes and humans.[17]

Where members of the English and German anthropological societies are concerned, the situation has not yet been fully analysed. They also discussed Darwinism, albeit to a lesser extent than their French counterparts in the *Société d'Anthropologie*. Erickson draws our attention to a moderately positive attitude towards evolutionary theory within the *Deutsche Gesellschaft für Anthropologie, Ethnologie und Urgeschichte*, and a very noncommittal reaction within the Anthropological Society of London. Rudolf Virchow, the foremost and most influential anthropologist in Germany, regarded the theory of evolution as an unproven hypothesis which only facts could substantiate, while in England James Hunt, founder of the Anthropological Society, was sceptical about Darwin's ideas, particularly in their application to Man.[18] Harvey suggests that not only in France but also elsewhere in Europe, a positivist and a materialist reaction to Darwin can be distinguished within communities of anthropologists.[19]

Despite Darwin's caution, with the publication of his *Origin of Species* he had ensured that the issue of human origins occupied the centre of the stage. Our special interest here lies in the role of fossils in discussions of human descent. This role seems obvious to the present-day reader, but to the nineteenth-century scientist the relevance of human fossils for the question of human origins was in no way self-evident. For an understanding of the background to this, we must first look at the way in which the human fossils known in the previous century were interpreted.

Thanks largely to Georges Cuvier's work at the beginning of the nineteenth century, the idea that the earth was extremely old, and that in the course of its history various animal species had died out, found acceptance within a wide circle. At first this did not affect the concept of the recent origin of Man. For most scientists, the human's position in the scheme of things was not a matter for discussion; the Bible taught that Man had only appeared on earth a few thousand years before, and there was no cause to call biblical authority into question.

Otherwise, it is not true that even Cuvier regarded humanity as God's last work of creation, or that he denied great antiquity to Man.[20] The source of this error is Cuvier's saying – often quoted but taken out of context – that "There are no human fossil bones."[21] By this Cuvier only meant that no human bones had been found in any of the fossil-bearing sediments known up to that time. However, he did not exclude the possibility that in times long past when the animals that later became extinct were still alive, humans may also have been living, even though not in the same regions as at present, but in remote corners of the earth.[22] Cuvier was silent on the subject of Man's origin; it was not his custom to raise questions to which he could produce no answers.

Nevertheless, Cuvier's cautious statements were interpreted by many of his contemporaries as a categorical denial of Man's extreme antiquity. The very idea of fossil humans was for them an impossible concept; humankind was of very recent origin and had certainly not inhabited the earth in the time of the now-extinct animals.

This view was to be gradually undermined during the course of the century. First, there were the finds. In the 1820s John MacEnery found stone tools in a Kent cave amidst the bones of extinct animals. At the same time Marcel de Serres, Paul Tournal, and Jules de Christol uncovered fossil human bones in France. The remains of prehistoric habitation sites were discovered on the banks of Swiss lakes, the first in 1829 and more in the 1850s, and prehistoric middens were found in Denmark. In the period 1829–1833 Philippe Charles Schmerling from Delft came across human and animal fossils in the Engis Caves near Liège.[23]

In this period an increasing number of scientists gradually decided that they should be guided only by their own observations, even when these did not tally with the biblical story. Thus the idea slowly gained ground that Man was older than earlier investigators had deemed possible. The breakthrough occurred about 1860 with the work of Jacques Boucher de Perthes, among others. Yet he had had to wait many years for recognition. His book *Antiquités Celtiques et antédiluviennes* of 1847 was initially consigned to virtual oblivion, although it must be said that the work's speculative bias, and the outdated theoretical context within which Boucher located his finds, were partly responsible for this lack of interest.[24]

Since 1841 Boucher had been collecting primitive tools close to where he lived in Abbeville; they were in a geological layer yielding up fossils of extinct elephants, rhinoceroses and bears, for example. His conclusion was that humans must have been living in the same period as these animals, and that Man was therefore considerably older than had generally been supposed. Similar finds made by Marcel Jérôme Riggolot at St. Acheul strengthened Boucher's conviction.[25]

Recognition of Boucher's work was given a particular boost by excavations in Brixham Cave in south-west England, where artefacts were discovered in association with fossil animal bones in the period 1858–1859.[26] Thanks largely to the meticulous precision with which these excavations were carried out, several English geologists of high prestige including Charles Lyell, Joseph Prestwich, and Hugh Falconer, were gradually convinced of the great antiquity of *Homo sapiens*. They also took a new interest in Boucher de Perthes' work and, after personal inspection of the artefacts and the layers in which they had been found, supported his conclusions. From that moment on Boucher's evidence for the great age of humanity attracted widespread attention. Recognition of his work also stimulated a renewal of interest in his predecessors' finds; Lyell discussed the nineteenth-century finds we have described in detail in his *Antiquity of Man* of 1863, for instance.

This, then, was the background to the discussion on fossil humans in the second half of the nineteenth century. There was a wealth of finds in this period, and by the end of the century many sites of varying degrees of importance were known. There is no need to deal with all the discoveries here; a discussion of the most important is sufficient to illustrate the way in which opinion on the significance of fossil human remains was formed.

Our first example is Neanderthal Man, discovered by Johann Carl Fuhlrott in 1856, in a cave in the Neander Valley near Düsseldorf (Fig. 1).[27] There were sharply divergent opinions over Fuhlrott's find, which consisted of a skullcap, two thigh bones, and a number of other skeleton fragments. Fuhlrott and Hermann Schaaffhausen, who published the first descriptions of the discovery, were convinced that the bones were extremely ancient, and to

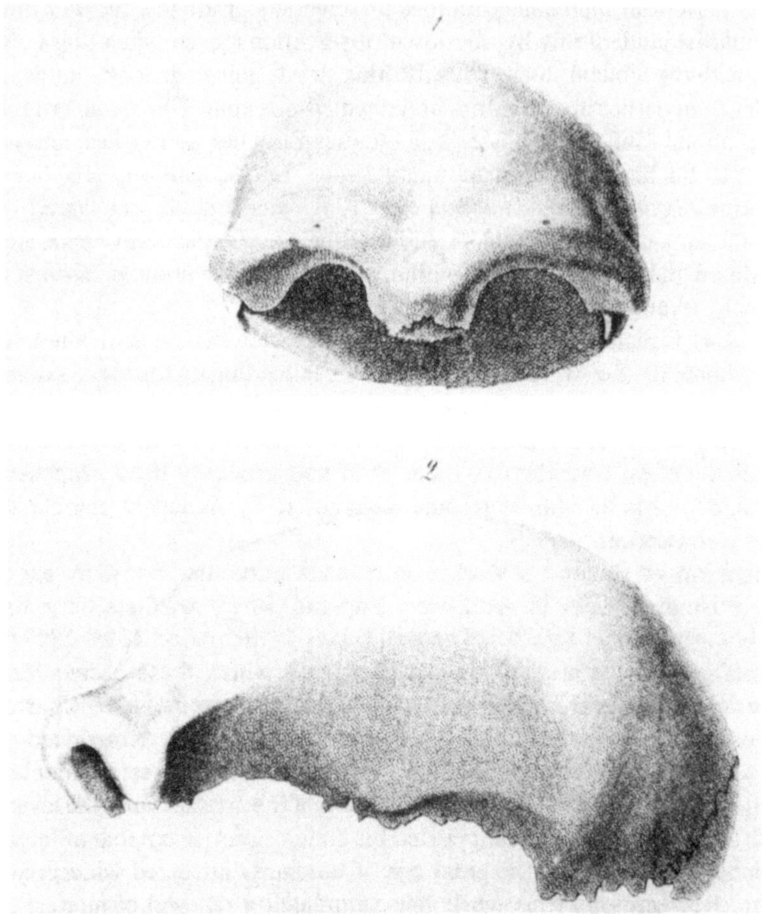

Fig. 1. The fossil skull found in the Neander Valley in 1856.

Schaaffhausen the features of the skull, for example the jutting eyebrow ridges, proved that they were dealing here with a hitherto unknown human race. Schaaffhausen suggested that Neanderthal Man had belonged to one of the 'savage' races mentioned by Roman writers, living in the north-west of Europe before the Celts and Teutons. He said nothing about the genealogical relationship between this race and modern humanity.[28] George Busk, who translated Schaaffhausen's article into English, thus initiating the discussion in Britain, pointed out several ape-like traits in the skull, but he did not address the question of descent either.[29]

Huxley discussed the Neanderthal bones in his *Man's Place in Nature* of 1863, observing that the skull showed a number of primitive features, but this is as far as he went. In his opinion there was no way in which the cranium could have come from a transitional form between apes and Man. Neanderthal Man belonged to a primitive but completely human race, comparable with the Australian Aboriginals.[30] Lyell, in his *Antiquity of Man*, adopted Huxley's conclusions without any modification,[31] while Darwin in contrast devoted barely any attention to the Neanderthal fossils, observing only that the skull was "well-developed and capacious," thereby emphasizing its human properties.[32] Wallace felt that there were no great disparities between the Neanderthal skull and that of the modern human.[33] Only the geologist William King saw a distinction great enough to place Neanderthal Man in a separate category, *Homo neanderthalensis*. In 1863 he described the species as more ape-like than any modern human race, but even he made no mention of the genealogical relationship between the Neanderthals and the contemporary human.[34]

According to Rudolf Virchow, the authority on pathology, the bones had been deformed by disease; he believed that he could detect traces of rickets and arthritis, for instance. Later on he suggested that the bones were those of a Frisian. The anthropologist Joseph Barnard Davis also argued that the skull shape ought to be regarded as an abnormality, and Richard Owen entirely agreed with him in this.[35]

Many anthropologists attributed the bones to a still-extant race. The possibility of a recent origin could indeed not be automatically excluded, since nothing certain was known about the age of the bones. Rudolph Wagner thought they belonged to an old Dutchman, while August Franz Mayer opted for a rickety Cossack, and Franz Pruner Bey for a Celt. Then there were others, such as Charles Carter Blake, who thought that the skull was that of an idiot.[36]

The conclusion reached by the materialists Büchner and Vogt was also negative: the bones offered no clues by which the origin of humankind could be documented.[37] Haeckel completely ignored Neanderthal Man at first; he only incorporated him into the human family tree in 1908, over fifty years after the discovery.[38] In their analysis of the Neanderthal bones, opponents of evolutionary theory such as Blake, Davis, and Pruner Bey observed that the fossils did nothing to make the idea of human evolution any more acceptable, yet this remark was somewhat superfluous since no one had so far claimed, in so many words, that they did so.

The story of two other significant finds, the first a Neanderthal jaw bone and the second several Cro-Magnon skulls and skeleton fragments, follows more or less the same lines. The jaw bone was discovered in 1866 by Edouard Dupont in the 'Trou de La Naulette' close to Dinant. More authorities acknowledged the great age of the find than had been the case with Fuhlrott's Neanderthal skull, since Dupont had also found the bones of extinct mammals in the Naulette cave. The fossil drew little attention in Germany or Britain, but Paul Broca in France initiated a discussion within the *Société d'Anthropologie* on the question of whether the jaw bone could serve as material evidence of human evolution.

Gabriel de Mortillet and Eugène Dally were inclined to believe that it could; in de Mortillet's view, the jaw bone displayed characteristics somewhere between those of apes and those of men. Pruner Bey, on the other hand, denied any significance to the issue of evolution; in his opinion, the jaw bone showed no essential difference from that of a member of a modern human race. Even Broca came to a negative conclusion, despite his readiness to look at the fossil from an evolutionary angle. True, it appeared to be the most primitive human jaw bone ever discovered, but in Broca's view it was premature to claim that it had belonged to a representative of some transitional form. General opinion inclined towards Pruner Bey's view that the jaw bone came from a member of a recent human race.[39]

The discovery of Cro-Magnon Man in 1868, near Les Eyzies in the Dordogne, produced little discussion on the question of human origins. In Broca's portrayal the Cro-Magnon belonged to a race superior to the Neanderthals but in certain respects more primitive than modern humans. Broca was silent on the subject of genealogical relationships. Pruner Bey stressed that the bone measurements fell within the scale of variation found among modern-human sizes. The question of evolutionary theory was not raised.[40]

Other finds of fossil human relics in the period up to 1885 offered no really new perspectives on the matter. The great age of the remains was sometimes acknowledged, sometimes doubted; nothing was said about transitional forms.

The role of palaeontology and anthropology

In the first section of this chapter we saw that the question of human descent was discussed very intensively in the years after the publication of Darwin's *Origin of Species*, yet human fossils did not play the role in the debate one would have expected. When both adherents and opponents took the available material into consideration, they concluded that it could not be accepted as evidence of human evolution. How can we explain this? Here we shall give both a general cause deriving from the significance attached to material palaeontological evidence, and a specific one connected with the theoretical and methodological context within which research into fossil hominids was carried out.

As we have established, for a great many years Haeckel ignored the Neanderthals, and expressed no views on the discoveries at La Naulette and Les Eyzies. His attitude was extreme yet not untypical, illustrating the lack of prominence given to palaeontology within the German school of morphology. Also, it was precisely this German school, which Haeckel had helped to found, that in the second half of the nineteenth century came to dominate research into the matter of human descent.

One of the chief aims of this school of thought within the area of morphology was the reconstruction of phylogeny, the evolutionary history of life. The morphologist had a choice of three research paths: comparative anatomy, embryology, and palaeontology. If the last named seems at first sight to be the most appropriate discipline for the purposes of phylogenetic reconstruction – after all, fossils provide us with the foundation stones for the sequence of descent – it was in fact comparative anatomy and embryology that predominated in Haeckel's work and in that of Carl Gegenbaur, the co-founder of the German school.[41]

This can be explained partly by the simple fact that, in practice, palaeontology came nowhere near to fulfilling the expectations that one could, in theory, have entertained about it. For this reason Darwin, in the formation of his theory of evolution, had drawn heavily on the familiar assumption about the imperfection of the geological record. There were also the objections to which, even today, palaeontology is unavoidably subject: that research material is so often fragmentary; that collection of the material is time consuming and difficult; and that such factors as chance and luck play a significant part here.

Yet the decisive factor for Haeckel was that he believed palaeontology to be unnecessary. Unlike Gegenbaur, who was never so dogmatic on this point, Haeckel considered the evolution of life to be an indisputable fact, requiring no further palaeontological proof. In his view, comparative anatomical and embryological research were adequate means for compiling the desired family trees. Haeckel's fundamental biogenetic law, that the ontogeny of an organism presents a condensed recapitulation of its phylogeny, took over palaeontology's role, so to speak. Supplementary evidence from the palaeontological side was welcome in support of the results obtained via embryology and comparative anatomy, but it was not actually necessary. Haeckel himself warned more than once against overestimating the value of palaeontological data. In discussing human origins he wrote, for instance:

> We can only partly acknowledge the great importance which laymen and narrow specialists attach to the evidence of such 'fossil men' and 'transitional forms between the apes and man'. He who has a thorough knowledge of comparative anatomy and ontogeny, as well as of palaeontology, and who is capable of an open-minded comparison of the phenomena, has no need of those fossil documents in order to accept the 'descent of man from the ape' as an historical fact. It appears to us that this is already... an empirically fully founded hypothesis, whether or not later palaeontological discoveries are going to uncover any 'intermediate forms'.[42]

Haeckel's viewpoint was extreme, but to a great extent Gegenbaur and the majority of morphologists also availed themselves of comparative-anatomical and embryological data for their phylogenetic studies, including investigations of human descent. Palaeontological research into our early human ancestors was not stimulated by the German school of morphology, which dominated the field of zoology until well into the 1880s.[43] All the discoveries of hominid remains described above, were accidental finds. No one mounted a deliberate search for the missing links between ape and Man, and when finds were actually made the reaction – if anyone did react – was an extremely reserved one.

As we have seen, those morphologists who did actually pronounce judgement on the subject of human fossils, usually arrived at negative conclusions, and the same applies to the anthropologists. To trace the causes for this negativity we must turn to nineteenth-century anthropology. The study of fossil humans constituted an overlap between morphological and anthropological working areas and both anthropologists and morphologists were unmistakably influenced by the theories and methods prevailing within the anthropology of the time.

From approximately 1860 on, the idea of the great antiquity of Man was more or less generally accepted, but for most anthropologists there was no connection between this idea and the subject of evolution. According to the consensus, there was no substantial morphological difference between prehistoric humans and their modern counterparts; the prehistoric human belonged at most to a separate race, while there was no concept of any direct genealogical kinship between prehistoric and modern races. The human fossils known at that time did nothing to prompt revision of this view.

Erickson has demonstrated convincingly that this position can be explained by a number of theoretical concepts employed within the field of anthropology, concepts originally having no connection at all with evolutionary theory, and dating from long before Darwin.[44] I shall deal here with the most important.

One of the main themes within anthropology was the question of the origin of the human races. There were two opposing views: the monogenist, in which all the human races had the same origin; and the polygenist, according to which each race was thought to have been formed or created separately from the others. In the eighteenth century the monogenist view was the most strongly represented, according as it did with the Genesis story. Yet in the course of the nineteenth century, in which the Bible was used less and less as the touchstone for the results of scientific research, the polygenist view gained ground. Later on we shall see why this was so. Toward the middle of the century it even gained the ascendency; prominent anthropologists such as Virchow, Hunt, and Broca were polygenists.

After the appearance of Darwin's *Origin* the anthropologists who accepted the theory of evolution continued the monogenism-polygenism debate in evolutionist terms, the monogenists defending the descent of the human races from one common ancestor, the polygenists accepting the idea of a separate derivation via a number of ancestral lines. In its most extreme form the polygenist perspective amounted to the view that the human races represented separate species, each of which had emerged from a different type of ape. Vogt, Büchner and Broca, for instance, held this kind of view. Darwin and Wallace inclined more towards the monogenist standpoint, although they also thought that the races had a long history of separate development. By no means all anthropologists acknowledged the validity of evolutionary theory, but even among the anti-evolutionists the monogenism-polygenism debate was carried on until the end of the nineteenth century. Hunt, for instance, was polygenist but opposed the application of evolutionary theory to Man. The prominent French anthropologist Armand de Quatrefages also rejected the theory, yet at the same time he defended the monogenist standpoint.[45]

Here we have outlined some of the background to the interpretation of fossil human remains. The polygenists, who were in the majority, saw no reason for assuming genealogical kinship between prehistoric and modern humans. The races had come into existence separately, each having its own origin, and this applied just as much to prehistoric as to present-day races. According to the polygenists there was absolutely no proof of the transformation of one race into another. Negroes remained Negroes even after generations of living under totally different circumstances to those in their ancestral countries, and Egyptians from the time of the Pharaohs were no different, morphologically speaking, from their modern counterparts.

Virchow, and many other positivistically inclined scientists with him, felt virtually obliged to accept the idea of separate origins for different races, on the one hand because proof of a common descent was lacking and on the other because, time and again, empirical research demonstrated great racial disparities.[46] This latter factor requires some clarification; we shall approach it via a second consideration that determined the context of anthropological interpretation.

Both polygenists and monogenists had taken up their positions long before the *Origin of Species* was published. Their conceptions of the origin of races were based on investigation of present-day peoples rather than on relevant data. Darwin's theory of evolution produced no reorientation within research; the study of extant races remained the basis for theories about their descent. In these investigations nineteenth-century anthropologists identified great racial disparities; however much they differed over racial origins, both monogenists and polygenists were agreed that there was a sharp distinction among the races. As we have described, some polygenists even talked of different species, but the monogenists, de Quatrefages for instance, while stressing both the changeability and potential equality of the races, perceived great disparities between them, too. It was usual for representatives of both positions to divide the races into a hierarchy, the 'superior' Caucasians at the top and 'savages' such as the Australian Aboriginals and Negroes at the bottom. Racist ideas were once the rule rather than the exception, and some even suggested that the differences between the apes and the most 'inferior' human races were narrower than those between the races on the bottom rung of the hierarchy and those at the top.[47]

This view of racial disparities largely stemmed from the markedly reductionist working method characteristic of nineteenth-century anthropology. In investigations of race, not only physical features and skin colour but also modes of behaviour and expressions of cultural life were traced back to inherent, inherited factors; all the emphasis lay on the 'nature' side of the 'nature-nurture' spectrum. Until well into the nineteenth century, the research showed a distinctly typological bias, with each race distinguished by a set of traits regarded as peculiar to that race. A standard, a type was assigned to each race, one could say.[48]

Craniometry or skull measurement, one of the most frequently used investigatory methods, was also based on reductionist premisses, great significance being ascribed to the skull's characteristics and measurements. The background to this was the assumption that the form and measurements of the skull reflected the construction and dimensions of the brain, and thus the intellectual, moral, and cultural capacities. Craniometric disparities between races were held to represent inherent qualitative difference.[49]

Anthropologists concentrated on the multiformity of races, rather than on the unity in diversity. In this respect the polygenists went further than the monogenists, but even the latter assumed the existence of wide racial disparities. The difference in opinion between the two groups was concerned exclusively with the implications of such disparities for the questions of both racial origins and the extent to which the disparities were fundamental and unbridgeable. Given these circumstances, it is hardly any wonder that the polygenists were in the majority and were to remain so for a considerable time after publication of the theory of evolution. When people believed that every race was an island in the ocean of humanity, so to speak, the idea of separate lines of descent was obviously taken for granted.

Both monogenists and polygenists thus assumed a wide racial spectrum with great and distinct differences between the races, and so with great diversity within the human species as a whole. This gives us a second explanation for the nineteenth-century interpretation of fossil remains; the prehistoric races known could be fitted with little effort into the broad range of modern races. Some saw the disparities between the Caucasian and Australian races, for instance, as greater than those between the Australian

Aboriginals and the Neanderthals. It was therefore unnecessary to do more than posit a distinct race for Neanderthal Man; and for some scholars even this was superfluous. Humanity was a conglomerate of very divergent types which could easily absorb the Neanderthals. Where a polygenist origin for the races was taken as a starting point, especially, there was absolutely no reason to pose questions of descent. For many, the problem of whether or not Neanderthal Man was a predecessor of modern Man simply never arose, and whoever did in fact raise the matter was met with a negative answer.[50]

For a third, and last, explanation we must turn to a nineteenth-century anthropological theory about the origins of the Caucasian race, a theory that induced both monogenists and polygenists to embrace the idea of a marked genealogical separation between the fossil human and the modern. At the beginning of the nineteenth century the idea spread that the ancestors of contemporary Europeans had not always lived in the West, but were the descendants of a people originating in Central Asia. In the first place this supposition was based upon linguistic data, in which structural similarities between the classic European languages and Indian Sanskrit had been identified, along with the confirmation of Sanskrit's greater antiquity. Language and race subsequently came to be identified with each other, and the conclusion was that the origins of European peoples had to be sought in India or Central Asia. Studies of old Indo-European races and cultures appeared to confirm this conclusion, and by mid-nineteenth century the idea of the Asiatic derivation of the Caucasian race had been raised to the status of virtual dogma.

The initial Indo-European thesis, also called the Aryan thesis, was subject to a good deal of adjustment and supplementation in the course of the nineteenth century, but all the later ramifications of the theory need not concern us here. What is essential is that firstly, until the 1880s the majority of anthropologists believed that contemporary European peoples ought to be regarded as immigrants rather than as indigenes; secondly, that the Asiatic invasion was visualised as having occurred at the end of the Palaeolithic or the beginning of the Neolithic. Given that the finds of fossil human remains in Europe dated from the Palaeolithic, a chasm between fossil and modern humans had opened up. The race from Asia was considered superior to the indigenous Europeans, the newcomers living by agriculture and livestock rearing, while the indigenous population were hunters. It was assumed that the inferior indigenous race was ousted from its customary living areas, and then virtually disappeared.[51]

The implications of the Aryan thesis for the interpretation of European fossil humans are clear: however Neanderthal Man, for example, was regarded, even if he was believed to be more primitive than any modern human, there could be no question of direct kinship with present-day Europeans. Caucasian racial roots lay in Asia rather than in Europe.

In conclusion we must draw attention to the role played here by Anders Retzius' skull-index theory, which distinguished between the long-skulled or dolichocephalic races, and the short-skulled or brachycephalic. Retzius suggested that the indigenous European peoples had brachycephalic skulls, while those of the Asiatic immigrants were dolichocephalic.[52] This explains why an anthropologist such as Pruner Bey believed that both Neanderthal and Cro-Magnon Man must have belonged to a modern race, since both had dolichocephalic skulls.[53] Conversely, we must observe here that a great many anthropologists regarded the dolichocephaly of these Palaeolithic races as in itself a refutation of Retzius' thesis on indigenous-European brachycephaly.

Erickson rightly disputes Stebbins' suggestion that anthropologists who adopted evolutionary theory incorporated it into their research.[54] Most anthropologists did in fact acknowledge the great antiquity of Man, but examination of fossil hominids did nothing to transform interpretation. A mixture of anthropological theories and methods was responsible for the fact that prehistoric human remains were not regarded as material evidence for human evolution. Lyell was even rapped over the knuckles by an anonymous reviewer of the *Antiquity of Man* because he had raised the subject of evolution in his book:

> that subject has nothing to do with the antiquity of man... No theory of development can be true without an enormous antiquity; but any amount of antiquity for the appearance of man or his works does not give any support to the theory of progressive transmutation.[55]

For others the reverse was true. For instance, from the great age of Man combined with the fact that prehistoric races could be fitted into the present-day racial spectrum, Wallace was able to derive an argument in support of his thesis that human evolution had come to an end. And the American Theodore Gill argued in 1874 that:

> the negative results afforded us indicate that fossil man was, in all respects, a typical man, perhaps even differing less from his successors in Europe than do some other existing races. It is at least certain that he had no decided ape-like characteristics... [T]he anxious may... contemplate with a happy serenity the explorations made, for every skeleton found, in its perfect manlike features, will not only disprove the existence of the dreaded intermediate link, but will add to the value of the negative evidence against such a link – that is in Europe or America.[56]

Gabriel de Mortillet was one of the few anthropologists who were inclined to explain morphological differences between prehistoric and modern humans in terms of transformation. The vast majority were not in agreement with this approach and acknowledged, at best, a taxonomic significance in the differences, certainly not a phylogenetic one. Biologists such as Huxley, Vogt, and Wallace were influenced by the then-current anthropological theories and methods in their interpretation of fossil hominids, and they also thought, for instance, that the Neanderthal fossils provided no evidence for human evolution. In their view the Neanderthals were a possibly new and primitive, yet fully human race, rather than our predecessors in the phylogenetic sense. We have also seen that many morphologists felt the palaeontological approach to the question of human descent to be unprofitable; Haeckel ignored even the fossils that were available. The generality of scientists devoting their attention to the question of descent were convinced that comparative-anatomical and embryological investigation was the appropriate way to document the history of human development.

Notes

1 See for example Herbert, "The Place of Man".
2 Darwin, *Origin*, 488.
3 See for example Zacharias, *The Construction of a Primate Order*, chapter 6; Desmond, *Archetypes and Ancestors*, 75–78, 143.
4 For Wallace's viewpoint on the position of Man, see Kottler, "Wallace, the Origin of Man and Spiritualism"; Schwartz, "Darwin, Wallace, and the Descent of Man".
5 Wallace, "On the Origin of Human Races".
6 Wallace, *Contributions*, 303–331.
7 See note 4.
8 For Lyell's view on human evolution, see Bartholomew, "Lyell and Evolution".
9 Huxley, *Man's Place in Nature*, 144.
10 See Zacharias, *The Construction of a Primate Order*, 204, 290–292; Montgomery, *Evolution and Darwinism*, 95.
11 Haeckel, *Natürliche Schöpfungsgeschichte*, 501–515.
12 Vogt published the lectures in 1863 under the title *Vorlesungen über den Menschen.* I have used the English translation, *Lectures on Man*, in which the microcephaly theory is found on p. 194–202. For this, see also Montgomery, *Evolution and Darwinism*, 96–97; Eiseley, *Darwin's Century*, 266–270.
13 See Montgomery, *Evolution and Darwinism*, 96–97.
14 Erickson, *Origins of Physical Anthropology*, 25.
15 For Darwin's reception in France, see Stebbins, *French Reactions to Darwin*; Farley, "Initial Reactions".
16 Harvey, "Evolutionism Transformed". See also Erickson, *Origins of Physical Anthropology*, 42–50.
17 Harvey, "Evolutionism Transformed", 292.
18 Erickson, *Origins of Physical Anthropology*, 39–42, 50–55. For the situation in England, see also Burrow, "Evolution and Anthropology"; Stewart, "The Effect of Darwin's Theory", 12–17.
19 Harvey, "Evolutionism Transformed", 310 note. I shall be dealing with the situation in the Netherlands in chapter two.

[20] See for instance Bourdier's article on Cuvier in the *Dictionary of Scientific Biography*. For an outstanding analysis of Cuvier's ideas on fossil Man, see Grayson, *Human Antiquity*, 46–54.

[21] Cuvier, *Discours*, 131.

[22] *Ibid.*, 138. Grayson (*Human Antiquity*, 50–52) posits that Cuvier tried particularly hard to show that human fossils were never encountered together with the remains of extinct mammals, in order to take the edge off Lamarck's contention that humans could have had a hand in either the vanishing of the fossil mammal species, or their transformation into modern species by domestication.

[23] For the finds described, and for similar discoveries in the same period and earlier, see Grayson, *Human Antiquity*; Lyon, "The Search for Fossil Man". See also the contemporary standard work *Antiquity of Man* by Lyell.

[24] Grayson, *Human Antiquity*, 120–132.

[25] *Ibid.*, 172–176.

[26] For these excavations and their importance for the acknowledgement of Boucher's work, see Gruber, "Brixham Cave"; Grayson, *Human Antiquity*, 179–195.

[27] For the discovery of Neanderthal Man and the subsequent discussion, see Gruber, "The Neanderthal Controversy"; Eiseley, *Darwin's Century*, 272–279; Erickson, *Origins of Physical Anthropology*, 140–151.

[28] Erickson (*ibid.*, 142–143) rightly states that, in Schaaffhausen's article, the idea of evolution was there in the background. As early as 1853 Schaaffhausen had shown evidence of thinking in terms of transformation in his "Ueber Beständigkeit und Umwandlung der Arten", but there is nothing to show that he saw Neanderthal Man as a human predecessor. It looks more as though he regarded the Neanderthals as a human race that died out precisely because of its primitiveness, while present-day races were descended from a less primitive form. Schaaffhausen ("On the Crania of the Most Ancient Races of Man", 159–160) wrote: "As we cannot now look upon the primitive world as representing a wholly different condition of things, from which no transition exists to the organic life of the present time, the designation of *fossil*, as applied to *a bone*, has no longer the sense it conveyed in the time of Cuvier. Sufficient grounds exist for the assumption that man coexisted with the animals found in the *diluvium*; and many a barbarous race may, before all historical time, have disappeared, together with the animals of the ancient world, whilst races whose organization is approved have continued the genus." (Translation by G. Busk; see note 29).

[29] Busk added his commentary to his translation of Schaaffhausen's article, which appeared under the title "On the Crania of the Most Ancient Races of Man".

[30] Huxley, *Man's Place in Nature*, 168–208, in particular 205–208.

[31] Lyell, *Antiquity of Man*, 75–92, 369. Assessment of the Neanderthal skull was hampered by the fact that one of the skulls found at Engis by Schmerling, which revealed a more modern morphology, was thought to be equally old.

[32] Darwin, *Descent of Man*, 146.

[33] Wallace, "On the Origin of Human Races", CLXVII.

[34] King, "The Neanderthal Skull".

[35] See note 27.

[36] *Idem.*

[37] Erickson, *Origins of Physical Anthropology*, 24–25.

[38] Haeckel, *Das Menschenproblem*, 27, 62.

[39] Erickson, *Origins of Physical Anthropology*, 155–159.

[40] *Ibid.*, 159–165.

[41] For concise but very clear discussions of Haeckel's and Gegenbaur's ideas, see the biographical essays by (respectively) Uschmann and Coleman in the *Dictionary of Scientific Biography*. See also Russell, *Form and Function*, chapter 14.

[42] Haeckel, *Systematische Phylogenie*, 618.

[43] Rudwick (*The Meaning of Fossils*, 261) describes it in this way: "The failure of the fossil record to provide positive evidence for evolution at the higher taxonomic levels caused many evolutionists to fall back on the less direct evidence of comparative embryology." Haeckel's biogenetic law "had the baneful effect of deluding biologists into believing they possessed a virtually infalliable [sic] key to the evolutionary ancestry of living organisms. In the euphoric atmosphere produced by the consequent re-interpretation of morphology and embryology, the fossil evidence for evolution was often ignored or, at best, gratuitously moulded to suit preconceived conclusions."

[44] Erickson, *Origins of Physical Anthropology*, 66–123, *passim*. Oldroyd gives a concise explanation for the scant influence evolutionary theory had on the formation of anthropological theory in his *Darwinian Impacts*, 298–308.

[45] Erickson, *Origins of Physical Anthropology*, 66–99; Stocking, "The Persistence of Polygenist Thought".

[46] See note 44.

[47] Examples can be found in virtually every discussion of nineteenth-century theories on race. Here we cite only Haller, *Outcasts from Evolution*.

[48] For the reductionist methodology and the typological bias of anthropology, see Erickson, *Origins of Physical Anthropology*, 77–81; Stocking, "The Persistence of Polygenist Thought"; Odom, "Generalizations on Race".

[49] See note 47.

[50] See note 44.

[51] Erickson, *Origins of Physical Anthropology*, 100–123. For the ideological background to, and implications of the Aryan thesis, see Poliakov, *The Aryan Myth*. After 1890 the Aryan thesis rapidly lost ground as an anthropological theory. The idea of an Asiatic origin was abandoned, but the term 'Aryan' continued to be fashionable. The theory, says Poliakov (*ibid.*, 327), "passed from the scientists to the demagogues to become at last the official doctrine of the Third Reich, when men designated as non Aryans were sacrificed to the gods of racialism."

[52] Erickson, *Origins of Physical Anthropology*, 114–115.

[53] *Ibid.*, 149, 161–165.

[54] Stebbins, *French Reactions to Darwin*, 203; Erickson, *Origins of Physical Anthropology*, 39–55, in particular 49–50.

[55] Quoted in Grayson, *Human Antiquity*, 211.

[56] Quoted in Eiseley, *Darwin's Century*, 276. For that matter Gill was no anti-evolutionist. He did not exclude the possibility that missing links might one day be discovered, but if so, they would be found in geological layers older than those in which the human fossils known up to then had been encountered.

Chapter 2

THE ROAD TO TRINIL

Eijsden and Roermond

Biographical sketches of Dubois quite frequently begin with three dates: 1856, discovery of Neanderthal Man; 1858, birth of Dubois; 1859, publication of Darwin's *Origin of Species*. It would not be entirely incorrect to suggest that here, in a nutshell, lies the best perspective for viewing Dubois' life and work, but it does not reveal the formative influences determining the course his life was to take. In this respect the fact that he was born in the South Limburg village of Eijsden carries greater weight.[1]

Dubois was a born scientist. He was passionately interested in nature study from childhood, and it was above all the Limburg landscape that stimulated his scholarly interests and gave them direction:

> I believe that the place where a man first lives is very important for his future... My birth place and first home... undoubtedly sowed the seeds of the naturalist in me, or at least developed them.[2]

His father, Eijsden's pharmacist, encouraged him in this interest. From an early age Eugène learned the names of the Limburg plants, and helped his father to collect medicinal herbs for the pharmacy. At secondary school he read Buffon's *Histoire naturelle*, and took handbooks such as Suringar's *Zakflora* and Calwer's *Käferbuch* on field trips. He also performed chemistry experiments, for which his father lent him his equipment. In this way he was offered every opportunity to indulge his scientific interests.

Dubois' later preference for palaeontology can also be traced directly back to his youth. The Netherlands is not richly endowed in the palaeontological sense, yet can nevertheless boast a long tradition of scientific investigation of its own territory: the study of the fossil-rich chalk deposits in and around St. Peter's mountain (*St. Pietersberg*) near Maastricht. As early as the eighteenth century both amateurs and professional scholars, Dutch and foreign, were examining the Maastricht chalk, subsequently creating rich collections and producing noteworthy publications. The spectacular investigation of the mosasaur (*Mosasaurus hoffmanni*) has been described many times, and all those interested in it will be familiar with the names of Drouin, Hoffmann, Godding, both the Campers, and Faujas de St. Fond. Thanks to Faujas we even have a monograph on the Pietersberg, an indication of the importance attached to this particular locality. Successive generations also built up valuable collections of less spectacular animal species. Bosquet, Binkhorst van de Binkhorst, Thierens, and Ubaghs are among those whose reputations largely rested on their treatments of the huge fossil wealth of the Limburg chalk.[3]

Granted that Dubois' name is not connected with research into the chalk, there was certainly a direct link between his interest in palaeontology and the area in which he was born:

> From my boyhood bedroom the continuation of the St. Pietersberg into Belgium [was] visible over the [river] Maas. I walked there many a time to collect fossils, and to the sand of the S[outh] L[imburg] chalk plateau close to the eastern boundary of Eijsden.[4]

Like most South Limburgers, Dubois' parents were Catholics. Considering that Dubois senior encouraged his son's inclinations towards science, we can assume that he was less mistrustful of the natural sciences than most of his co-religionists.[5] There is of course nothing strange in a pharmacist behaving in this way; he looked forward to a pharmacist's career for his son, as well. Evidently, he also attached great importance to good training in the natural sciences since, after Eugène had finished primary school, his father sent him to the State HBS (*Hogere Burger School*, or Dutch High School), an institution viewed askance in Catholic circles; according to a leading Catholic politician, the school produced "nothing but intellectual and moral ruins."[6]

The foundations of Dubois' later career in the natural sciences were laid during his time at the HBS:

> the HBS at Roermond... was probably more important to me than the university... I believe that if I had attended a grammar school (*gymnasium*), much of my aptitude for natural science would have been lost.[7] The Roermond HBS was something in the nature of a small university for me... blessed with superb teachers and just as well equipped as the universities of that time, [with] a first-rate chemistry laboratory, a large zoological, mineralogical, and petrographic collection, [and] a fine botanical garden.[8]

The teachers who, according to Dubois, contributed most to his development were Jos. Rosen, a student of the German chemist and geologist C.F. Mohr; W.H. Julius, later to become Professor of Physics at Utrecht; and J.M.A. Kramps, who taught mathematics and geology.

AMSTERDAM

"My father wanted me to be a pharmacist, but I was drawn more to medicine."[9] Medical studies won the day and in 1877 Dubois left for Amsterdam to begin them.[10]

There is only scanty information available about Dubois' student days. He often spent time within a circle of the prominent Catholics of that time, including the publisher and man of letters J.A. Alberdingk Thijm – aptly named "mentor... to all Catholics who wanted to study"[11] – the architect P.J.H. Cuypers and the rhetorician, commentator, and priest J.W. Brouwers. Clearly, in this period Dubois' qualities must have been recognised, since within these circles he was regarded as a very promising young man of talent.[12]

In spite of these Catholic connections it was in these years that Dubois finally turned his back on the Catholic faith. Later in life he said that, as early as his secondary-school days, he had distanced himself from the faith largely because it impeded his thinking; he claimed that he wanted to be completely free in this respect.[13] As far as I know Dubois never expressed any atheistic ideas, but he did sometimes show evidence of fiercely anti-Catholic sentiments.[14] His attitude towards religious belief as such can best be characterised as agnostic. In Dubois' view faith and science ought to be kept separate, and his publications confirm that he held strictly to this principle.

Very soon after beginning his studies Dubois lost any aspirations he might have had to become a practising physician.[15] The chief interest of his youth, nature study, maintained its place in the forefront of his life, even in Amsterdam. Where the natural sciences were concerned – the first year of medical studies was entirely devoted to them – Dubois could not have wished for a better training ground. Within a short time the teachers appointed to the faculty of natural sciences were to take the university forward into a period of great achievement. The names of Van der Waals, Van 't Hoff and De Vries, speak for themselves. Within the medical faculty the physiologist Thomas Place and the morphologist Max Fürbringer were the teachers who contributed most to Dubois' education.

In 1881 Fürbringer appointed Dubois as his assistant. This was the start of a successful career as an anatomist, but Dubois did not demonstrate absolute conviction in his choice for anatomy. Immediately after Fürbringer had offered him an assistantship, Place did the same, and Dubois was to

Fig. 2. Dubois in 1883.

remark later that: "If [Place] had asked me a day earlier than Fürbringer, I would have become a physiologist, which attracted me more." He added the rather singular statement: "Then I would never have gone to the [Dutch East] Indies."[16] The true value of these remarks, made years after the event, is naturally difficult to gauge, yet one implication seems clear, at least: Dubois was not unconditionally dedicated to anatomy.

From the information available it is impossible to deduce with any certainty why he felt more attracted to physiology. It may well be that his interest in this subject was connected with his admiration for Thomas Place, an admiration bordering on hero-worship.[17] Dubois was not alone in this veneration; Place was an inspired teacher who was idolised by his students. Yet his teaching was far from being up to date. Barely any practical or experimental work was carried out, the emphasis lying rather on theory and criticism.[18] In view of this, it seems rather unlikely that Dubois' affinity for physiology stemmed from a preference for experimental rather than descriptive research. Moreover, it is impossible to detect any such preference in Dubois' works or unpublished papers. If there were actually any question of Place influencing Dubois, it was through the stimulation of his interest in a theoretical approach to physiological problems such as those Dubois tackled in some of his later publications.

At all events anatomy it was to be, rather than physiology. Dubois' teacher Fürbringer, trained by Haeckel and Gegenbaur at Jena, was an out-and-out morphologist and a worthy exponent of the German school of morphology. After years of being assistant to Gegenbaur he was appointed to the post at Amsterdam, after a vain attempt had been made to entice Gegenbaur himself. For Fürbringer as much as for Gegenbaur, comparative anatomy took pride of place; his magnum opus *Untersuchungen zur Morphologie und Systematik der Vögel* is telling evidence of this.[19]

As Fürbringer's assistant Dubois was given a comprehensive introduction to morphological research, working also on comparative anatomy. There is clear evidence that he was regarded as a gifted student. When Fürbringer was offered the post of lecturer in anatomy in the State Training College for Teachers of Art (*Rijksnormaalschool voor Teekenonderwijzers*) and the State School for the Applied Arts (*Rijksschool voor Kunstnijverheid*) in 1881, he successfully recommended Dubois in his place. In 1885 Dubois was invited to take up a Readership in anatomy at Utrecht, with F.C. Donders personally going to some trouble to persuade Dubois. However, after consulting Fürbringer he declined the offer, which suggests that Fürbringer was able to hold out even more favourable prospects to him.[20] In a letter of 1891 Dubois wrote about this: "I and others expected that, after the departure of Prof. F[ürbringer] (which could not be long delayed because of [his] continual suffering from the Dutch climate) I was going to take over his post."[21]

In about 1885 Dubois carried out his first independent piece of research, a study of the comparative anatomy of the larynx in vertebrates. Fürbringer probably suggested this topic, since he had also worked on the problem, and had thus spotted many gaps in knowledge of the anatomy of the larynx.[22] Dubois' research was in complete accordance with traditional phylogenetic morphology. Its aim was to produce a careful comparative-anatomical description of the larynx in various animal species, to be followed by interpretation of the data in terms of the evolutionary history of the organ.

Dubois' main achievement in this study was to establish that the thyroid cartilage, part of the larynx, was homologous to the fourth and fifth branchial arches. He published a summary of his results in 1886 in the *Anatomischer Anzeiger.*[23] Besides this, he wrote the section on the larynx in whales for Max Weber's *Beitrag zur Frage nach dem Ursprung der Cetaceen.*[24] Then he made preparations for a comprehensive publication on his larynx research. He was making steady progress with the manuscript, the drawings were being made – and then Dubois threw it all up. On 29 October 1887 he set out for the Dutch East Indies on the steamship *Prinses Amalia* to investigate our human ancestors.

To the Dutch East Indies

Why this sudden volte-face? Up to now, barely any thought has been given to the circumstances surrounding Dubois' decision to go to the Dutch East Indies, although we are concerned here with the crucial change of direction in his career. At this point a sketch of these circumstances is desirable.

I have to begin with one reservation: I have no answer to one of the most intriguing questions, about the reason why Dubois rather than anyone else felt called to go in search of the missing link. The sources make it possible to comprehend why he gave up his career in Amsterdam. His arguments for choosing the Dutch East Indies for his journey of discovery are also easy to reconstruct precisely. Yet how he came to take such a step, and why he was prepared to burn all his boats behind him in so doing, continues to be a riddle. There are a few indications that the issue of human descent interested him, but there it ends, and the riddle remains unsolved. There are no 'external factors' that could be taken as decisive, and unless a hitherto unknown source ever comes to light which can tell us more about Dubois' internal motives, we shall continue to be in the dark as to the driving force behind his decision.

With this reservation in mind let us now turn to the context within which Dubois' plan was formulated. Above all, his departure for the East Indies implied an escape from Amsterdam. He could not stand the work in the anatomy laboratory, and his tasks and duties weighed heavily on him – especially teaching, for which Dubois had an outspoken and lifelong loathing. Later in life, when he was more or less compelled to face the student benches once more, he would often escape to the dunes after his classes, to recover from the nightmare of teaching.[25]

As we have already seen, Dubois had not opted entirely wholeheartedly for anatomy, and evidently the time came when he had to pay the price for this lack of commitment. This can be clearly seen in the following quotation from a letter to Dubois written by the zoologist Max Weber, giving his opinion over Dubois' intention of going to the East Indies.

> After thinking more about your plans it has become even clearer to me... that whether or not you go to the East Indies, you have lost your link with anatomy... From your view of teaching and its value and from your pessimistic ideas of science and from your outspoken dislike of teaching I think that for you, happiness can never lie in theoretical explanation. Posts of this kind are so hard to come by that they ought not to be occupied by people who are not completely contented in this working environment... You know the work demanded in anatomy and you didn't like it, unless it involved your own research; you liked it so little that you wanted to choose another working environment. I think that you have already decided this matter in your heart, have already made a decision, regardless of whether you go to the East Indies or establish yourself here as a doctor.[26]

Judging from Weber's letter Dubois still had some interest in his own research but it seems doubtful, to say the least, that he was prepared to sell his soul for it. He was an unusually restless person, and restlessness is incompatible with the great patience, precision, and dedication required for laboratory work. Dubois was good at producing intuitive ideas and rapid generalisations, but often lacked the interest and patience for working out his data and for constructing, step by step, the firm basis and argumentation in support of his conclusions. One of his students wrote that:

> Dubois had the habit of just lifting a corner of the veil of a scientific concept, but he was loath to settle down and work it out thoroughly; he preferred to leave to others with more perseverance the task of continuing and finishing it.[27]

Something that happened between him and Fürbringer also played an important part in Dubois' final decision to part company with anatomy.[28] However trivial the reason for this unpleasantness, its consequences were far-reaching.

What exactly happened? When Dubois brought the manuscript of his first larynx study to Fürbringer, the latter commented that he had established long before that the human thyroid cartilage is homologous with the fourth branchial arch, and claimed that he had been drawing attention to this in his lectures since 1880. Dubois thought that this was not a valid claim, but said nothing. On the contrary, he gave Fürbringer a free hand in altering the manuscript in such a way that it incorporated his claim. Dubois soon came to regret this because, as he was to say years later, Fürbringer had swindled him out of his discovery.[29] Yet this was probably not the real reason for his regrets. Whether or not Fürbringer's claim was a valid one, it is unlikely that he intended to claim priority in the matter. In the first place his comment concerned only the human thyroid cartilage, whereas Dubois' conclusions applied to mammals in general. Also, Fürbringer was talking about homology with the fourth branchial arch alone, while Dubois showed that both fourth and fifth arches were involved. Lastly, Fürbringer's idea was not even new, because Wilhelm His had already suggested the same thing, and in print rather than just verbally; Dubois even mentioned His' article in his own publication.[30] There was thus probably no question of anyone having prior claim to the discovery.

What must really have upset Dubois was that Fürbringer's addition to the article could give the impression that Dubois had not made his discovery in an entirely independent way. He was afraid that Fürbringer would be seen as the instigator and intellectual parent of the research, and Dubois himself only as the industrious disciple carrying out the task assigned to him by his mentor. Dubois was hypersensitive on this matter.[31] A significant number of the many 'incidents' in his long career can be traced to his excessive concern for what he called his intellectual property rights, because of which he harboured a deep mistrust for his colleagues, being continually afraid that others might gain the credit for his work. Once he suspected anyone of attempting this, his reaction was often quite disproportionately fierce.

This was the case in the Fürbringer incident, and slowly but surely Dubois completely lost his trust in Fürbringer, who was unaware of what was happening and consequently had no idea at first of Dubois' growing mistrust. Ultimately he was to be presented with the bill for his 'transgression'. After Dubois' decision to go to the East Indies, Fürbringer offered him to take on the task of editing the comprehensive larynx study.[32] This offer would seem proof enough that he was unconscious of having done any harm to Dubois. It was clear then that Dubois would not get down to the work before his departure. From Fürbringer's letters to Dubois it is evident that a great deal of work was still needed to prepare the manuscript for printing, and it was therefore a magnanimous offer. What is more, Fürbringer demanded no share in the publication; the whole thing was to be published under Dubois' name alone. Yet from the moment Fürbringer made his proposal, he saw no more of Dubois and received either no answers to his letters, or only noncommittal ones, in which the offer was not mentioned at all. He was not even given the opportunity to see Dubois off when he left for the East Indies, since his question about the ship's departure was simply left unanswered. He received a farewell letter from Dubois only after the ship had sailed. The publication of the larynx study never materialised.[33]

What happened is clear: Dubois had abandoned his belief in Fürbringer, and scholarly cooperation with him had become impossible. This was a significant factor in his decision to turn his back on anatomy. He said later: "This was one of the secondary motives, apart from the main one, the search for the missing link, that decided me to leave for the East Indies..."[34]

Dubois' reasons for going to the East Indies were thus partly negative ones: he wanted to get away from Amsterdam. Let us now see how much the sources can tell us about the circumstances surrounding his main motive for leaving.

Later in life, Dubois recounted how he had been captivated, back in his Roermond days, by the discussion over human origins. He cited the lectures given in 1868 by Carl Vogt in the Netherlands; he had not attended these himself, but they had been well spoken of at the HBS.[35] Vogt's radical-materialist ideas acted as a catalyst in Dutch discussion of Darwinism. Darwin's theory had elicited only a few chance controversies until Vogt's visit, but afterwards discussion flared up in Dutch newspapers and periodicals.[36] It is quite likely that Dubois was first confronted with the 'ape theory' while at the HBS, through his acquaintance with Vogt's ideas, but it is only later on, in 1886, that we find hard evidence of his interest in the question.

Dubois must have been deeply involved with general evolutionary theory in his student days, since it was part of the theoretical basis for morphology, the discipline in which he qualified under Fürbringer's guidance. Besides Fürbringer, Dubois' teachers De Vries and Place must also have stimulated his interest in the question of evolution. It is clear from Place's inaugural speech as new professor at the Amsterdam Atheneum that he supported the theory and had a marked interest in it, since his speech was devoted entirely to the theory of descent.[37] Dubois had wide-ranging talks with De Vries on questions of origins,[38] and while in Amsterdam he immersed himself in the works of Darwin, Wallace, Lyell and Haeckel.

Dubois must have derived his knowledge of research into fossil humans mainly from foreign literature, since in the Netherlands barely anything had been done in this area. The only Dutch find of any significance was a lower jaw that had come to light during the excavation of a canal between Maastricht and Hocht, from 1815 to 1823. Some nineteenth-century scientists assumed the piece to be very old, but later it was confirmed that the jaw bone was of recent date.[39]

Dutch research in physical anthropology in the second half of the nineteenth century concentrated on the question of the racial origin and composition of the Netherlands' population. Craniological studies predominated, and data were derived from local populations or from grave finds of the recent past,[40] while research into fossil humans was not discussed.[41] In the Netherlands there was no question of large-scale archaeological research into the culture of prehistoric people, such as that being carried out in France for example. Palaeolithic artefact-bearing strata or caves were unknown. The subjects of archaeological research dated either from the Neolithic, as for instance the megalithic tombs (*hunebedden*) and the Limburg flint mines, or from an even more recent period, as with the *terpen* (mounds on which habitations were raised above land subject to flooding) and remains from

Roman times. Attention was thus devoted entirely to the manifestations of recent human life.[42]

One of the objects of study mentioned above, the Limburg flint industry, did stimulate Dubois to undertake his first investigation of the prehistoric. In 1881 the Belgian archaeologist Marcel de Puydt discovered the Neolithic flint mines close to Rijckholt and St. Geertruid. On the basis of De Puydt's finds Dubois went to the vicinity of Rijckholt in 1886, to search for human and animal remains. Here he came across a geological 'organ pipe' known as the Henkeput and set up an investigation there in 1886 and 1887 during his vacations, assisted by the Count De Geloes of Eijsden, who owned the territory in which the Henkeput was located. The results were disappointing. Certainly large quantities of bones were found, including those of dogs, cattle, foxes, and deer, and there was a great number of human bones as well, but these were all of recent origin. In view of the few pieces of Roman earthenware also encountered, Dubois posited that the human remains had been flung into the Henkeput after a battle in Roman times. Neither prehistoric human remains nor artefacts were ever found.[43]

The Henkeput excavations provide the first – and as far as his Amsterdam period is concerned, the only – concrete evidence of Dubois' interest in the question of human ancestry. For the rest, the only thing of which we can be certain is that the problem fascinated him to the extent that, in 1887, after it had become obvious that his future did not lie with anatomy, he made the decision to devote himself entirely to the quest for the missing links between ape and Man.

One aspect of this decision deserving of special attention is Dubois' choice of a palaeontological approach. The background to this was partly of a personal nature. As we have already seen, his interest in palaeontology dated back to his youth, and we can safely assume that he was more attracted to field work than to laboratory research. He was a typical outdoor type, and a great nature lover. Yet he had another reason for his choice which he later outlined as follows:

> Embryology and comparative anatomy and atavisms can only provide indirect proof for the existence of the close tie between Man and the animal world, only palaeontology provides direct proof.[44]

Here Dubois was consciously breaking away from the tradition of research into human ancestry, as we have portrayed it in the previous chapter. It is obvious that he did not share Haeckel's viewpoint that palaeontological evidence for human evolution was superfluous. He also distanced himself from the indirect approach, based on study of modern primates, which had predominated within morphology up to that time. According to Dubois, evolution ought to be documented on the basis of material proofs of the actual process of development. As we shall show, he thus chose an approach that was to alter the face of palaeoanthropology drastically.

What were Dubois' grounds for trying his luck in the Dutch East Indies? Both practical and theoretical concerns played their part in this. Turning to the practical aspects first: the research had to be financed. Dubois applied to the Dutch government for subsidisation, but drew a blank.[45] This refusal was not altogether surprising since it was obvious to everyone, including Dubois himself, that his chances of success were extremely small.[46] Furthermore, the nature of the investigation was controversial, to say the least. Add to this the fact that the Netherlands was in the grip of an economic depression, at its worst in about 1885, so that financing for private scientific projects was in any case not high on the list of priorities, and the government's reaction is entirely understandable.

As a result Dubois had to ensure that he would be able to maintain himself. Since he believed that the missing link had to be sought in the tropics, the Dutch East Indies offered the most obvious opportunity; at least the road to the East Indies lay wide open to the Dutch. Dubois followed in the footsteps of such naturalists as Pieter Bleeker and Cornelis Swaving[47] and joined the Royal Dutch East Indies Army (*Koninklijk Nederlandsch-Indisch Leger*) as medical officer, second class.

Dubois also had other arguments, involving the content of research, for selecting the East Indies in preference to other tropical regions, for instance Africa or British India. This much is clear from an article he published in the *Natuurkundig tijdschrift* in 1888 while stationed on Sumatra, entitled "Over de wenschelijkheid van een onderzoek naar de diluviale fauna van Ned. Indië, in het bijzonder van Sumatra" (On the Advisability of Research on the Diluvial Fauna of the Dutch Indies, Especially of Sumatra). In this he put forward five arguments for assuming that the East Indies was the most appropriate region for investigating human predecessors. We shall now deal with these arguments in logical order, regardless of the order in which Dubois presented them, returning later to the structure of his reasoning, which had been carefully weighed.

Dubois derived his first argument from Darwin, who in his *Descent of Man* had reasoned that our distant human ancestors must have lived in the tropics, since human beings had lost their fur pelts in the course of their development. The climate in the temperate zones was considerably colder in the Pleistocene than it is today, so that human development could not have occurred there.[48]

According to Dubois this last hypothesis was confirmed by the human fossils encountered in the European Pleistocene strata. It is clear from his comment that, in spite of hacking out a new path through the problem of fossil humans, he was still indisputably influenced by the ideas about fossil and modern human races that were current in the second half of the nineteenth century. He suggested that:

> Even allowing that the famous Neanderthal skull – more specimens of the same type were found a short while ago in Belgium – and the lower jaw from La Naulette and the one from the Schipka Cave are really anything other than pathologica... then they are nothing more than proof of the existence in Europe, during the diluvial period, of low-level yet still never really pithecoid races, not even markedly inferior to the lower races of today.[49]

Here we see that Dubois believed that Neanderthal Man, even if he had belonged to a distinct race, could easily be accommodated within the broad range of modern races, and thus did not deserve a separate status.

This view is all the more striking when we consider that the Neanderthal fossils discovered in 1886 near Spy in Belgium by Marcel de Puydt and Max Lohest, to which Dubois was referring in the quotation given above, offered every opportunity for a reinterpretation. Lohest and Julien Fraipont had published an excellent description of the Spy bones.[50] There could be no doubt about the great age of the bones; they displayed the same features as the La Naulette jaw and the skull from Neanderthal, virtually confirming that the characteristics were racial in nature rather than due to accidental defects. What was even more important was that the authors were clearly inclined to locate the Neanderthals in an evolutionist framework, thus distancing themselves from the traditional interpretational context. They believed that there was an enormous chasm between ape and Neanderthal Man, and that more ape-like forms must have preceded the latter, but they also argued "that the ethnic type of lower-Quaternary Man has travelled a very long road to reach [the stage of] the present types if he is [indeed] the source of [these types]."[51] Nevertheless, Lohest and Fraipont were given no support, and Dubois was certainly not the only one to turn a deaf ear to the new song they were singing.[52] The implications for the idea of human evolution embodied in the find were to be recognised only at the end of the century.

For Dubois, the presence of Neanderthals in the European Pleistocene offered only a prop for the argument that the forbears of humankind had to be sought in the tropics. Humans had arrived in Europe only after the process of humanisation had already been completed.

As his second argument for looking in the East Indies for the missing links between ape and Man, Dubois pointed out that the apes, our closest relatives in the animal world, were only found in the tropical regions of the Old World. He reasoned that in every area where the climate had not been subject to radical change since the Pleistocene, present-day animal species were directly descended from the fauna that had inhabited the area during the Pleistocene. This also applied to the tropics, and one could thus expect to find the immediate predecessors of the apes in tropical regions. Granted that both humans and apes had developed out of a common ancestor, it was self-evident that our far-distant ancestors had also lived in the tropics. This restricted the probabilities for human genesis to Africa with its gorilla and chimpanzee, and South-East Asia where the orang-utan and gibbon are found.[53]

Darwin, Dubois noted, had suggested Africa as the most likely place for the birth of humankind, arguing that Man was the closest relative to the chimpanzee and gorilla, but Lyell and Wallace had left the possibility of the East Indies open.[54] This brought Dubois to his third argument – now directed more especially towards the East Indies as the cradle of humanity – in which he appealed to material palaeontological evidence. Referring to Darwin's, Wallace's, and Lyell's views, he observed: "Yet none of these three great Englishmen was then familiar with the Siwalik chimpanzee, which makes it seem to us more probable that the East Indies was the last habitation of the prehistoric Anthropoids."[55]

This Siwalik chimpanzee had been discovered in 1878 in Pliocene sedimentary deposits in the Siwalik Hills, in the present-day border area between India and Pakistan. The find, consisting only of an incomplete upper jaw with a number of teeth, was documented by Richard Lydekker in 1879, first under the name of *Palaeopithecus sivalensis*;[56] in 1886 he changed the name to *Troglodytes sivalensis*.[57] According to Lydekker the jaw fragment came from an anthropoid that could be regarded as a predecessor of the chimpanzee. Besides this chimpanzee, the canine tooth of a fossil orang-utan had also been found in the Siwaliks, and Lydekker deduced from both finds that the late-Tertiary ancestors of the apes had very probably lived in oriental regions.[58]

Dubois felt that the Siwalik-chimpanzee find was of great significance for the Dutch East Indies investigation as well. He declared that, although the palaeontology of the Indies had as yet hardly been studied, there were clear signs that the Pliocene faunas of Borneo, Sumatra, and Java corresponded with the Siwalik fauna. Since Java's fossil vertebrates, documented by the Leiden geologist Karl Martin,[59] were so similar to those from British India, he concluded that the Indo-Malayan islands had been joined to the mainland in earlier times. Wallace had come to the same conclusion on zoogeographical grounds; in his view the connection with the mainland had been broken only in fairly recent times, geologically speaking. Because of this both the Pliocene and Pleistocene fauna Siwalik Hills would have been able to disperse into the Indo-Malayan region. The migration to the Dutch East Indies, and also to Africa, must have been prompted by the cooling climate in the Siwaliks at the beginning of the Pleistocene, resulting from the Ice Age; the tropical fauna must have moved away to the south, south-east, and south-west in that period.[60]

Since fossil primates had been discovered in British India, Dubois believed that there were justifiable hopes of similar finds in the Dutch East Indies, and where the predecessors of the apes had been found, one could also expect to find the forbears of Man.

For his fourth argument Dubois directed attention to the occurrence of gibbons in Further India and the Indo-Malayan archipelago. In his view there were indications of a close kinship between human and gibbon. The Miocene fossil apes *Pliopithecus* and *Dryopithecus*, the last of which was closer to Man than any living anthropoid, both showed resemblances to the gibbons.[61] He provided no concrete arguments to back up this statement, nor did he mention the sources from which he had drawn it – he had never actually seen the fossils in question – but there can be no doubt that he based it on the descriptions by Edouard Lartet and Alfred Gaudry. Lartet had classified *Pliopithecus* (discovered in the south of France in 1837) with the gibbons[62] and had attributed gibbon-like features to *Dryopithecus* (found in the same region in 1856) as well as similarities to humans.[63] Gaudry had defined *Dryopithecus* as the most humanlike of all the apes then known.[64]

In Dubois' opinion Lydekker's *Troglodytes sivalensis* from the Siwaliks fitted into this pattern. Granted, Lydekker had portrayed the animal as a predecessor of the chimpanzees, but he had also pointed out its resemblances to both gibbon and human.[65] Dubois concluded from this that "the Hylobates group which in earlier geological periods had developed further" could throw light on the history of human descent.[66] British India and the Indo-Malayan archipelago were the right places in which to investigate the accuracy of this proposition, since it was only there that gibbons were found.

Dubois' fifth and last argument for choosing the East Indies clearly had its roots in the traditions of nineteenth-century research into human fossils. He argued that all the major discoveries of fossil human remains had been made in caves, and he consequently expected that the exploration of caves would offer the best chance of success. From this viewpoint the Dutch East Indies were superbly suited to scientific investigation of this kind. There were caves in abundance on the Indo-Malayan islands, especially on Sumatra where, Dubois noted, there was a mountain range in the Padang Highlands called *Boekit Ngalau Sariboe*, 'the mountains of a thousand caves'. He thought that the mountain limestones of Sumatra were better fitted to cave exploration than the much softer limestone of Java. Experience had taught that there were far fewer caves in soft calcareous rocks.[67]

So much for Dubois' reasoning. All things considered, his choice of the Dutch East Indies was well thought out and well argued. He employed mainly theoretical arguments drawn from authoritative literature on the subject, but also indicated the palaeontological information available. It was especially Lydekker's discovery of a fossil ape in the Siwaliks and the fact that this, along with *Dryopithecus* and *Pliopithecus*, shared features in common with the gibbons, that persuaded Dubois to opt for the East Indies as his research area.

Dubois derived his theoretical arguments from views that were common property in his time: his ideas on the Neanderthals and his preference for the tropics and for cave excavation plainly reflect the nineteenth-century consensus on these. His palaeontological approach to the problem was, however, completely new. Up to that moment all the known human fossils had either been discovered purely by chance, or had been found by scientific investigators such as archaeologists and geologists working with no explicit intention of digging up the extinct predecessors of Man, or of finding proof for the theory of human evolution. Dubois was the first to base a quest for fossil humanity on a fully argued research plan.

We can still profit from further elucidation of several points from Dubois' argumentation. In the Dubois Collection there is the manuscript of a lecture Dubois gave at the beginning of 1888, probably to his colleagues in the army medical corps on Sumatra.[68] It dealt with the same subject as the article we have just discussed, but in the lecture Dubois was less circumspect in the way he put forward his ideas; on several points he showed more of his hand to his listeners than he did to his readers.

To begin with, it is more obvious from his lecture than from his article that he was a confirmed Darwinist. He even began the lecture with a hymn of praise to the *Origin of Species*. "This book brought about a reorientation of our entire knowledge of Nature, it was the signpost to a new era and a new world view." Darwin, Dubois continued, had put paid to the teleological view of Nature, and only his doctrine of natural selection could satisfactorily explain the development of life forms. Humankind had no claim to a unique position:

> nothing gives [Man] the right to demand an exception, a privilege for himself, to lay claim to a special act of creation for his appearance in the ranks of living beings... How high... the ants stand above most other insects, because of the physiological properties of their central nervous system; yet is that any reason for setting them apart from the others?

The question of when the transition from ape to human had taken place was dealt with in only a cursory way in Dubois' article, but in his lecture he made his views plain. He considered it unlikely that humans had lived as early as the Tertiary, as some maintained: "Everything which has been advanced as proof for this, such as cut marks in bones from Tertiary layers, has not been able to withstand stringent criticism."[69] The first humans lived in the Pleistocene, he continued, and taking this as his starting point, he reasoned that:

> The human type must have come into being rather rapidly. The first anthropoid with greater intellectual capacities than others of his own kind must have had a great advantage over them, and this ability to think more deeply must have been well preserved and rapidly increased in the struggle for existence. The human cannot therefore have appeared very much earlier than the Pleistocene, it is more likely that he came into being in the Pliocene age...

Dubois also disclosed to his audience something of what he hoped to find in the East Indies:

> although the human being has been found nearly the whole world over in a fairly highly developed state in the diluvium, it is nevertheless still very possible that in the same period intermediate forms were still living in the tropical forests of the Malay archipelago, which had not really earned the name of human... [They] could have survived for a long time, given the necessary isolation.

Finally Dubois pointed out, as he had done in his article, that since several fossil apes showed human as well as gibbon-like features, an expedition to the areas where the gibbons now live held out promising prospects for the search after our forbears:

> [On the basis of] the continuing divergence in characteristics resulting from the struggle for existence [one may] entertain some hope that, even in the Pleistocene, the human was already coming to the fore in some species, just as in some others [it was] the chimpanzee-like and the singular characteristics of the long-armed apes [i.e., the gibbons] – but less than in the similar species of recent times.

In conclusion it is worth mentioning a remark Dubois made at the end of his lecture on the Spy find of 1886, saying that this discovery made it plausible that the Neanderthals' characteristics were those of a race rather than accidental defects. In the following chapter we will hear him arguing the exact opposite.

SUMATRA

On 11 December 1887 Dubois arrived in Padang on Sumatra to take up his appointment to a hospital post (Fig. 3). For the first few months he was fully occupied with his hospital work, and had barely any time to spare for scientific research. At his own request he was seconded in May 1888 to

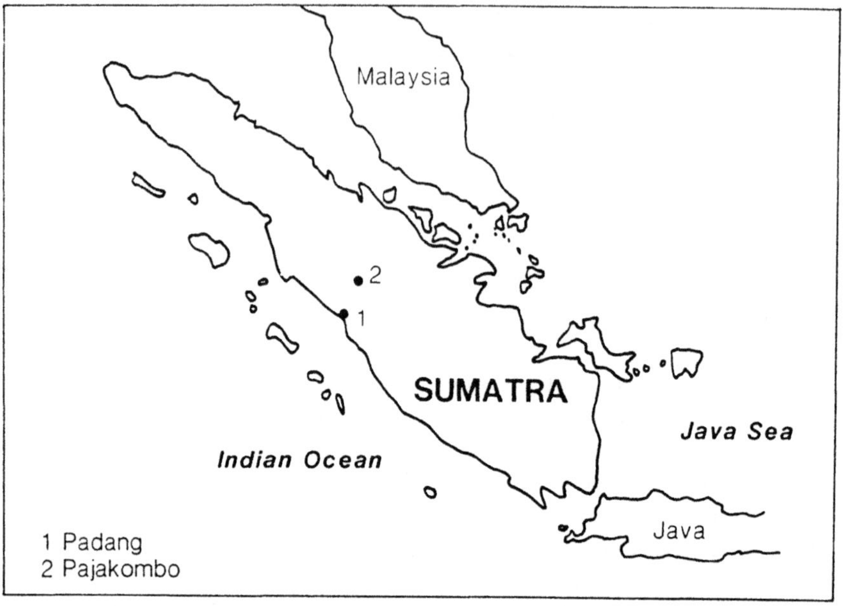

Fig. 3. Sumatra

Pajakombo[70] where he was put in charge of a hospital for army convalescents. Here he had more opportunities for pursuing his scientific work. A colleague who was recuperating there took some of Dubois' medical duties off his hands, enabling him to begin his search in the vicinity of Pajakombo for caves suited to palaeontological excavation.

In the meantime, in April 1888 Dubois had written his article on the desirability of palaeontological investigation in the Dutch East Indies, and the piece was published in the same year. The chief aim of the article was, naturally, to interest the Dutch East Indies Government in this kind of research project, and to obtain governmental financing for it. The structure of the article seems to have been determined mainly by tactical considerations. Out of a total of eighteen pages, the first eleven were devoted exclusively to presenting the favourable prospects for palaeontological study of the Dutch East Indies fossil fauna. Only in the last seven pages did Dubois discuss the possibilities of research into fossil humans. Thus even though some might consider the quest for human ancestors to be a hopeless venture, there was always the virtual certainty of good results from an investigation of the fossil fauna.

Dubois' argument also appealed to feelings about national prestige. He pointed out that there was great interest abroad in archaeological and palaeontological research, while the Dutch East Indies were in this respect practically virgin territory, even though it was precisely the East Indies which held out the greatest promise. "[The] time cannot be far off" he wrote "that foreign scholars will also turn their eyes towards the Dutch East Indies."[71] He closed with the rhetorical question:

> And will the Netherlands, which has done so much for the natural sciences of the East Indies colonies, remain indifferent where such important questions are concerned, while the road to their solution has been shown?[72]

Dubois hoped that the investigation he was carrying out on his own initiative from his base at Pajakombo would show that his article had not exaggerated matters. After an initial exploration to familiarise himself with the area, he began the actual investigation in July 1888. By August he had already had some success, in the Lida Adjer Cave. The animals he unearthed there included orang-utans, gibbons, rhinoceroses, tapirs, elephants, deer, cattle, and pigs. Here was the proof that fossils were actually to be found in the caves of Sumatra.[73]

The article in the *Natuurkundig tijdschrift* soon attracted attention. Dubois' ideas gained support from the East Indies Committee for Scientific Research (*Indisch Comité van Wetenschappelijk Onderzoek*),[74] especially from its chairman Melchior Treub; from the influential geologist Rogier D.M. Verbeek; from his superiors in the army medical corps; and from several highly placed government officials including W.P. Groeneveldt, Director of the Department of Education, Religion and Industry, and R.C. Kroesen, Governor of Sumatra's West Coast. Meanwhile, in the Netherlands the geologist Karl Martin and the zoologist Max Weber and the members of the Committee for the Promotion of Research in the Natural Sciences in the Dutch Colonies (*Commissie tot Bevordering van het Natuurkundig Onderzoek der Nederlandsche Koloniën*)[75] among others, were pleading Dubois' cause before the government. Lastly, Dubois himself wrote a report on the results achieved up to that time, and this was placed before the Dutch East Indies Government.[76]

Dubois was now in a much stronger position than he had been when he asked the Dutch government for a subsidy. He enjoyed the support of scientists and government officials; moreover, a good deal of emphasis was being placed on research into the fauna, which was certain to be rewarding to judge by the first results.

For once his plans were successful.[77] In a government resolution of 6 March 1889 he was seconded to the entourage of the Director of Education, Religion, and Industry (ERI) to carry out palaeontological research on Sumatra and possibly on Java as well. Two members of the engineering corps, Franke and Van den Nesse, were assigned to his party along with fifty forced labourers to help him with the excavation work.

Dubois now no longer confined his search to the vicinity of Pajakombo, but extended his working area over the whole of the Padang Highlands. However, positive and successful as the investigation of the first cave – the Lida Adjer – had been, subsequent excavations were difficult and disappointing. The work went sadly awry and the results were far poorer than his high-flown hopes had led him to expect. For an impression of the setbacks Dubois suffered we cannot do better than allow him to tell us in his own words about what went wrong. On 17 October 1889 he wrote in a letter to F.A. Jentink, Director of the National Museum of Natural History (*Rijksmuseum van Natuurlijke Historie*) at Leiden:

> Everything here has gone against me, and even with the utmost effort on my part, I have not achieved a hundredth part of what I had visualised. Where the cave explorations are concerned, the reverses began right at the start, with my coming here in the *poeaza* (the period of fasting) when the Malays are as indolent as frogs in winter... A survey of the caves I was... provided with seemed fitted only to put me off the scent [because] there were very few real caves among them. Yet these did exist, as I saw later, and people had simply concealed their existence from me... because they thought that the 'Company' would appropriate the gold and saltpetre the inlanders get out of these caves... After these experiences I went searching without guides and in this way I have found a few very useful caves, but still never the best one could wish for. What's more, it was necessary to live out in the forest for weeks on end, usually under an overhanging rock or in an improvised hut, and it turns out that in the long run I can't stand up to that, however well I was able to bear the fatigue [at first]. Having now come back, with my third bout of high fever, which nearly put paid to all the searching for 'diluvialia', I have to give it up for good.[78]... The trouble with personnel was even worse. To begin with one of the two engineers assigned to me to supervise the forced labourers, was totally useless, and after repeated warnings and exhortations to carry out his duties properly he was transferred at my request... Meanwhile the other engineer died from fever... The number [of forced labourers] placed at my disposal is 50. Of these, some (7) have run away or have been turned out because of misbehaviour; at present (in the rains) 50% are sick. There are also foremen and cooks [among them] so that at the moment – it is sad, but true – 12–15 labourers are [actually] working. A third drawback lies in the site itself. [The fact] that this is overgrown with forest would not be so bad in itself, if there were only roads or at least paths, and if the steep limestone mountains did not make communication with most points as good as impossible, and if there were not a total lack of water in many of the mountain places.

Yet Dubois did discover a few good find-spots, such as the Sibrambang and Djamboe Caves close to Tapisello, in which the fossil fauna showed obvious similarities to those of the Lida Adjer.[79] One disadvantage in all the Sumatra find-spots was that they usually produced only teeth, bones seldom being discovered. Dubois posited that the bones had been gnawed away by porcupines; in his opinion the gnaw marks on the few bones that had been preserved, and a trial with a live porcupine, were sufficient proof of this.[80]

By far the greatest blow, however, was the relatively recent date of the fauna unearthed. Dubois seemed to have been mistaken in his prediction that an extinct ancestral fauna would be discovered in the Sumatran caves. The bones found were certainly fossilised and were not of recent date, but Dubois believed that almost all of them came from still-extant species. He reserved judgement on their age in his reports but later, when he was working on Java, he dated the fauna to the Holocene.[81] Yet the excavations did reveal that the geographical distribution of the fauna had been different from that in modern times. The orang-utan, for example, had obviously existed in greater numbers and had occupied a wider territory on Sumatra.[82]

Since there was no difference, or barely any, between the fossil fauna and that of the present day, there appeared to be considerably less chance of finding forms that were transitional between ape and Man in the caves. Dubois was always careful to avoid this conclusion in his reports, originally perhaps because he had not abandoned hopes of unearthing older finds, but later probably entirely in order to disguise the fact that the Sumatran investigation had been a fiasco in this respect.

At the end of 1889 Dubois decided to continue his excavations on Java; on 17 October of that year he wrote to the geologist Verbeek, who had already advised him to go to Java:

> I had already abandoned my original grounds for preferring Sumatra when I heard from you about the very many places in Java where Tertiary limestone is found that can match the Sumatran carboniferous limestone for hardness. In this Java offers unmistakable advantages, in the far easier means of communication and in the infinitely greater extent of cultivation than in the regions here.[83]

Moreover, Martin had published a description of a Pliocene Javanese fauna, related to the Siwalik fauna of British India, thus confirming Java's possession of a fossil fauna that was not identical with the animals living on the island today.

An extraordinary find made in Java on 24 October 1888 provided Dubois with an added argument for transferring the excavations to Java. The mining engineer B.D. van Rietschoten had discovered a fossil human skull in a rock shelter close to the hamlet of Wadjak in East Java, while looking for marble

(Fig. 4). Van Rietschoten sent the skull to C.P. Sluiter, the curator of the Royal East Indies Society of Natural Science (*Koninklijke Natuurkundige Vereeniging*) at Batavia, for further examination.[84] Sluiter, however, did not consider himself qualified to give an opinion on the skull, and forwarded it to Dubois,[85] who was absolutely delighted at the discovery, and told Sluiter:

> In every respect... the skull differs so greatly from the type of the present-day inhabitants of the West-Malay islands [that] there can be absolutely no doubt that it has nothing to do with that present-day race... While I don't dare to make a definite statement before I have had the opportunity of classifying it precisely through close study and comparison, the skull seems to me to show the greatest similarities with the Papuan type, and I am virtually certain that the first representative of the primordial people of *Java* has now been discovered.[86]

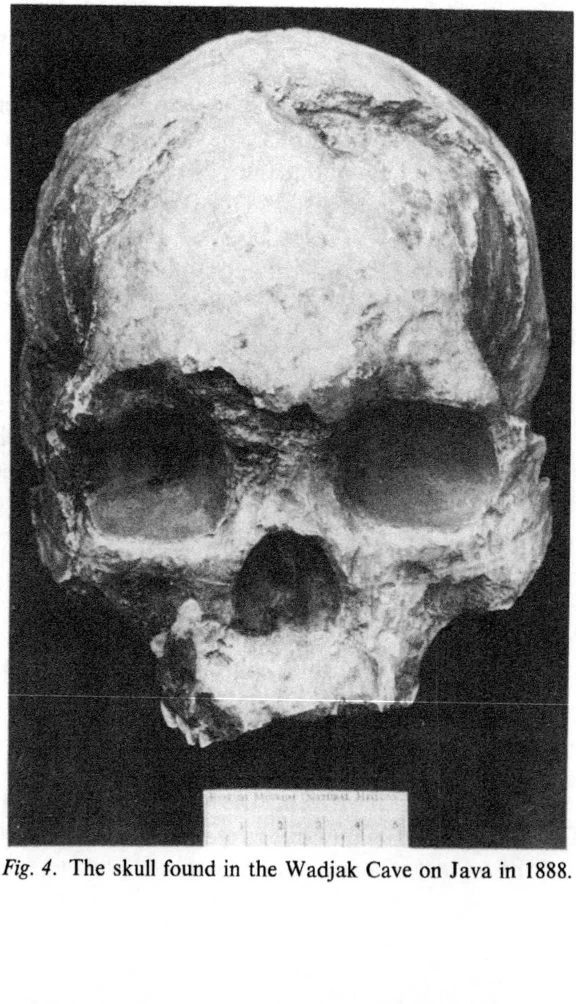

Fig. 4. The skull found in the Wadjak Cave on Java in 1888.

Dubois said nothing about the age of the skull, but it was obviously fossilised and so he could be sure that it did not date from the recent past.

Dubois saw Wadjak Man as a member of a human race (as he did in the case of the Neanderthals) rather than as a predecessor – in the evolutionary sense – of modern humans, a point to which I shall return later on. Nevertheless, the discovery proved that fossil human bones could be found on Java. The prospects for unearthing the forerunners of Man from the island's older geological strata, thus seemed better, the more so since the existence of these older layers had been confirmed, along with their fossil wealth.

The government responded sympathetically to Dubois' request for a transfer, and in a government resolution of 14 April 1890 gave him permission to extend his investigations to the Tertiary and Pleistocene sediments on Java.

JAVA

On Java the search was resumed in June 1890, Dubois' first trip being to the limestone mountains in the Madioen and Kediri Residencies, where Van Rietschoten had found the Wadjak skull (Fig. 5). Various caves were combed thoroughly in the following months, without any significant results except in the case of Van Rietschoten's rock shelter. A first exploration produced only a few fragmentary mammal bones, difficult to classify, but towards the end of September a fresh examination uncovered a new skull, not as complete as the first but unmistakably from a member of the same race. Later on, several fragments of a Wadjak Man skeleton were also found in the cave.[87] In one of the progress reports that Dubois produced at stipulated

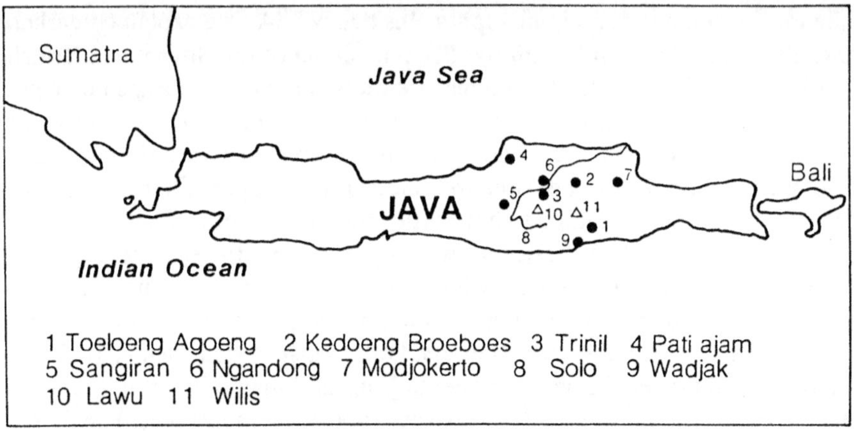

Fig. 5. Java.

times for the attention of the ERI Director, he observed a propos of these finds that:

> Further examination of the human remains shows that these most probably come from representatives of the Australian race, and therefore suggest a significantly higher level of development than those from La Naulette, Schipka, or Spy. These Australian inhabitants of Java undoubtedly lived in a time one could call geologically recent.[88]

From this it is clearer than ever that Dubois considered Wadjak Man to be on a par with modern human races; for him the find had no direct relevance to the question of human descent.

It has always been rather surprising that Dubois only published a comprehensive description of Wadjak Man some thirty years after the discovery. This long delay can now be at least partially explained. Wadjak Man originally held no significance for Dubois as a transitional form, and the description of the find was consequently not high on his list of priorities. He was to wait until 1910 before removing the hard sediment still partly covering the fossils, and scrutinising the bones closely. Around that time, however, his views on the status of the Neanderthals and comparable fossil forms underwent a radical change. As we shall see in chapter 4, Dubois consequently came to view the Wadjak finds in a new light.

Dubois had begun his search on Java in the traditional way, with cave exploration, but he now decided to carry out excavations in open territory as well, and this must be regarded as a decisive step. It had been known for a long time that fossil mammals were also to be found outside the caves on Java; various investigators had mentioned the fact. The best known, the Javanese painter Radèn Saléh and the German naturalist Franz Junghuhn, had uncovered fossils in the Kendeng Hills in Middle and East Java and on the Pati Ajam, a mountain in Japara Residency.[89] As we explained earlier, Martin linked these finds with the Pliocene fauna of the Siwaliks in British India.[90] Dubois now also turned his attention to them, encouraged by news from Verbeek, among others, that there were countless fossils still to be found in those regions. He began by dividing his time between the caves and the open field, but later he came to concentrate on open-site explorations, when the cave work seemed to have little more to offer of any importance.[91]

Dubois had now been assigned two corporals from the engineering corps – G. Kriele and A. De Winter – who were more than equal to their task and who carried out the excavations, sometimes together and at other times separately, each with half the labourers. Dubois installed himself at Toeloeng Agoeng, making regular visits to the diggings and maintaining written contact with the engineers over the way the work was progressing. The fossils unearthed were sent to Toeloeng Agoeng, where Dubois prepared them and made provisional identifications.[92]

The field exploration began at Kedoeng Loemboe, where Radèn Saléh had already been successful in collecting fossils, and at nearby Kedoeng Broeboes. From September 1890 other places in the Kendeng Hills were also being worked. The result was an unprecedentedly rich haul. In his third quarterly report of 1890 Dubois informed the ERI Director:

> From the western to the eastern end of the mountain range – that means almost the entire distance from *Semarang* to *Soerabaja* – and where the southern slope is concerned, also over the whole width of the range, we have found representatives of [the Siwalik] fauna everywhere, while they are also known to exist in several spots on the northern slope, which has not as yet been examined more closely. Thus the *Javanese Siwaliks* are in no way inferior in extent to those south of the *Himalaya* range... One really has only begun to get to know this fauna, and the *Javanese Siwalik formation* also bids fair to match the wealth of the British Indian [formation].[93]

Find-spots, unusually rich in fossils, were also discovered in the banks of the Solo River. The fossil mammals excavated in the Kendeng area and beside the Solo included cattle, deer, a primitive elephant (*Stegodon*) and a more modern one (*Elephas*), also hippopotami, rhinoceroses, pigs, hyenas, and felines. Crocodiles, river turtles, and freshwater molluscs also came to light. The fauna contained various extinct species and genera, plus several forms still in existence but nowadays found only outside Java. Dubois deduced from this material that the sediment from which the bones had been taken was a fluvial deposit, and that the finds must date from the Pliocene since they were similar to those from the Siwalik Hills.[94]

The splendid harvest from these sites decided Dubois to concentrate on the excavations there, and work was carried out in these regions in 1891, 1892 and 1893. However, digging could not be carried on uninterrupted in any one place, since the waters of the Solo rose and flooded the find-spots in the river banks during the monsoon, in the last months of each year. The disadvantage of the Kendeng Hills, on the other hand, was the fact that here exploration was hampered during the dry season by a dense layer of *djati* (teak) leaves covering the soil. The excavation work was therefore always arranged to permit digging at the Solo River in the dry season, until the onset of the rains made it impossible to continue, whereupon the exploration was shifted to the Kendeng Hills, where the same rains had in the meantime washed away the *djati* leaves, rendering the localities fully accessible once more. Broadly speaking this meant that the work went ahead in the Kendeng from December until May, and at the Solo River for the rest of the year.

Several other find-spots were explored, but Dubois felt that they yielded less than the Kendeng and Solo sites. In March 1891 an investigation was begun on the Pati Ajam in Japara Residency, where Junghuhn had discovered fossil mammal bones back in 1857. According to Dubois, the fauna that came to light here was identical with that of the Kendeng, even to the kind of sediment in which the fossils had lain. Yet there were fewer finds, and these were inferior in quality, so the Pati Ajam diggings were closed down in August 1891.[95]

In September 1893 find-spots at Tandjoeng and Sangiran were investigated, with similar results – a fauna akin to the Kendeng type, but not so fine in quality. In 1893 a short exploration of sites in the western part of the Buitenzorg and the eastern part of the Lebak departments produced nothing.[96] Attention therefore continued to be focussed on the Kendeng Hills and the Solo River; only the scope of the project was widened by looking into the outlying fossil-rich Kendeng strata beyond the hills.

What results were there to justify this exclusive attention? Dubois' starting point was the supposition that he was dealing with a Siwalik fauna, but by the last months of 1890 he could announce an important new insight into the palaeontology of Java. In his monthly report of November 1890 he argued that:

> Closer study of this fauna makes it increasingly probable that it is geologically younger than the British-Indian Siwalik fauna with which Prof. Martin compares it, and to which I too, until recently, thought it belonged.

The species discovered on Java were clearly not identical with those of the Siwaliks, but were more closely linked with the present-day faunas of Java and the mainland. Dubois came to the conclusion that the fossil fauna occupied an intermediate position between that of the Siwaliks and that of today, most probably dating from the Pleistocene.[97]

In his view the Javanese fauna was of a continental kind, which tallied with Wallace's supposition, based on zoogeographical data, that Java had still been joined to the continent of Asia in the Pleistocene. The fossils had been unearthed from sediments that were volcanic in origin and consequently Dubois thought that a volcanic eruption must have been responsible for the death of the animals concerned.[98] In 1892 he studied the structure of the sediments to determine the main direction from which the volcanic materials had flowed. His conclusion was that the eruption had probably been in the Wilis volcano.[99]

Since the fossils were situated in a fluvial sediment, Dubois surmised that the bodies of the animals dying in the eruption were washed down to the river by heavy rains. The fact that most of the bones were broken and were only very seldom encountered in their natural relation to each other, he attributed to animal violence. There were no traces of land-animal teeth on the bones, but there were in fact round holes that could have been caused by the cone-shaped teeth of the crocodile.[100]

The excavations increasingly reinforced Dubois' conviction that the fauna – which he regarded as an homogeneous whole – was more recent than that of the Siwaliks. Although he had originally opted for an early-Pleistocene age for the fossils, in 1893 he realised that a late-Pliocene dating was also possible. He wrote to Verbeek on 10 June 1894 that:

> at any rate until nearly a year ago... along with the English geologists and palaeontologists I thought the Siwalik fauna came from the Pliocene, then when I studied what has been published recently on the late-Tertiary mammal faunas in Europe (of Pikermi Samos, etc.) and especially after reading Zittel's views on these, I have gradually come to realise that [the Siwalik fauna] is predominantly late-Miocene, as its similarities to the Miocene faunae of Europe prove.

This implied that the Javanese fauna could also be older, possibly dating from the late Pliocene.

In 1894 Dubois looked into several geological aspects of the find-spots, determining for instance the position and extent of the bone-bearing strata, their thickness and inclination, and the direction in which they lay. He regarded this investigation as confirmation of his earlier conclusions on the age of the Javanese fauna and the cause of its destruction.[101]

Back in January 1892 Dubois and the Director of the ERI department had together drawn up plans for publishing the results.[102] Dubois thought at this point that he would be able to write up the greater part of the description within a year, but it soon became obvious that he had grossly underestimated the amount of work needed. The lack of literature and of comprehensive comparative collections was clearly a major handicap. Material that had to be ordered from the Netherlands and British India often took a long time to arrive, and there was a steadily growing collection of fossils waiting to be classified. Moreover, the East Indies climate was bad for Dubois' health, making it impossible for him to devote himself fully to the writing-up.[103]

Gradually it became clear to him that he would not be able to finish his task in the Dutch East Indies, and that he had to return to Europe. Before he did finally leave in the summer of 1895 he made a trip to British India early in the year, to study the collection of Siwalik vertebrates at the Indian Museum in Calcutta. At the same time he visited several of the localities in the Siwalik Hills, collecting a number of fossils there, including a molar from *Stegolophodon stegodontoides*, an extraordinary find considering how rare the fossil remains of this species are – right up to the present day only a few are known to have been found.[104] This visit caused no radical alteration in Dubois' vision of the relationship between the Siwalik and Javanese faunas.

If what we have discussed above were all that reflected the results of Dubois' excavations, his work could still be called successful, but the most significant discoveries have not yet been mentioned. These deserve to be dealt with separately.

Notes

[1] The biographical information on Dubois in this chapter has been derived (unless otherwise given) from the publications mentioned in note 2 of the Introduction.

[2] The quotation is taken from an unpaginated text with autobiographical notes, held in the Dubois Collection. There are two such texts, both bearing the caption "28–1–'28". The notes they contain were intended for Dubois' speech surrendering his professorial post. From here on these texts will be referred to as Autobiographical Notes.

[3] For the mosasaur research, see Van Regteren Altena, "Achttiende-eeuwse verzamelaars" and "Nieuwe gegevens"; Theunissen, "A.G. Camper". For the nineteenth-century Limburg palaeontologists, see Van den Bosch, "Van Breda en de Commissie voor de Geologische Kaart".

[4] Autobiographical Notes.

[5] For the Dutch Catholic community's attitude towards education and science in the second half of the last century, see Brom, *Herleving van de wetenschap*; Smit, "De houding van het katholieke volksdeel"; De Valk, "De katholieken en het onderwijs". According to De Valk (*ibid.*, 741) there was a small upper stratum of liberal Catholics who were less strict in enforcing the prohibition, exacted by Pius IX in the encyclical *Quanta Cura*, on education that was "divorced from Catholic belief and the power of the church."

[6] Brom, *Herleving van de wetenschap*, 111–112. The politician quoted is Herman Schaepman.

[7] Anon., "Prof. Dubois blikt over zijn leven terug".

[8] Autobiographical Notes.

[9] See note 7.

[10] The medical practitioners' laws of 1865 gave HBS pupils the means of sailing round the reef of the supplementary state examination in the classics, in order to make an immediate start on university medical studies. This meant, however, losing the chance of gaining an academic degree. HBS pupils could take the medical examination but they could not do a doctoral degree. Dubois did not take the final state examination and therefore had no right to take a doctorate (L.D. Brongersma, personal communication).

[11] Brom, *Herleving van de wetenschap*, 67.

[12] This is clear from, for instance, the fact that the Cuypers considered him a suitable candidate for marriage to their daughter Mia (Bernsen, Journal I, 19 November 1930). Mia Cuypers, however, married a Chinese government official, F.G. Taen-Err-Toung, in 1886. At the time she was making the acquaintance of this man (to her parent's dismay), Dubois, so he said, had to act as a 'lightning conductor'. "I had rather a good appearance and was expected to have a brilliant future" (Bernsen, Journal I, 19 november 1930), he added without a trace of false modesty. It was clearly of no avail, though. The story of Mia Cuypers' marriage, much talked of at the time, was used by man of letters Lodewijk van Deyssel (Mia's cousin) as the springboard for his novel *Blank en geel* (White and Yellow) published in 1892. My attention was drawn to this by the authority on Van Deyssel, Harry G.M. Prick, who edited a reissue of *Blank en geel* in 1979. At the same time he pointed out a name-change in Bernsen's report of the events. I have pleasure in extending my thanks to him for his help.

[13] Bernsen, Journal I, 10 November 1930; III, 25 September 1931, 19 November 1931. In a letter to J. Boeke, dated March 3, 1932, Dubois wrote "I have... to explain, that since my 13th year I have fallen away from the Catholic Church and have never again been under its influence for a single minute, right up to the present, neither in feeling nor deed."

[14] On 26 February 1933 he wrote, also to Boeke: "I [shall maintain] that Catholicism is the worst thing... the human race has ever encountered..."

[15] Later, as a medical officer in the Dutch East Indies, he found that medical practice had even begun to pall, as we can see for example from a letter from F.A. Jentink to Dubois on 7 June 1888, talking of Dubois' "distaste for military service and medical practice." To Bernsen (Journal II, 19 February 1931) he said: "I am always described as a doctor. But I'm an anatomist."

[16] Bernsen, Journal I, 17 December 1930.

[17] In illustration of this, I quote a letter from Dubois to Place, dated 13 October 1896: "In the East Indies, because of leaving Amsterdam, thereby sacrificing such a lot, one of the heaviest burdens I had to bear was the thought that I would be judged less favourably by men whose opinion I valued. Yet the thought that kept my spirits up, and the question I put to myself every day, was: What would Place think of me?... That you are convinced that I have done my best makes me much happier than I was... I have always believed in you completely and steadfastly, and my great admiration for you has increased over time. Nobody could admire you more than I do, truly."

[18] For Place, see C.A. Pekelharing, "In memoriam Thomas Place"; Van Rijnberk, "Prof Place"; Brugmans, *Gedenkboek*, 191, 654–655.

[19] For Fürbringer, see Bluntschli, "Max Fürbringer"; Uschmann, *Geschichte der Zoologie*, 68, *passim*; Brugmans, *Gedenkboek*, 189–190, 587.

[20] All this can be gleaned from the following letters: Dubois to Donders, 24 December 1885; Donders to Dubois, 28 December 1885; the curators of Utrecht University to Dubois, 29 December 1885.

[21] Letter to R.C. Kroesen, 18 January 1891. Fürbringer left for Jena in 1888.

[22] Fürbringer, *Zur Kenntniss der Kehlkopfmuskulatur*.

[23] Dubois, "Zur Morphologie des Larynx", 1886.

[24] Dubois, "Ueber den Larynx", 1886.

[25] Personal communication from F.H. van der Maarel, one of Dubois' old students.

[26] The letter is not dated, but was most probably written in 1887.

[27] Van der Feen and Van Benthem Jutting, "Dr. Antje Schreuder", 122. W.S.S. van Benthem Jutting is an old student of Dubois'.

[28] Bernsen described the incident in his Journal I, 29 January 1931.

[29] *Idem*.

[30] Dubois, "Zur Morphologie des Larynx", 1886, 180.

[31] To give another example from the same period: we can gather, from a letter of Fürbringer's to Dubois on 2 October 1886, that Dubois did not much like the idea of Fürbringer being at his first official lecture in his new post. He felt that Fürbringer's presence would give the impression that his old mentor was still supervising his work. "So the question arises,"Fürbringer chaffed, "whether I am at least permitted to attend the lecture with a muzzle, or whether you are afraid that even this might damage your independent status."

[32] Letter from Fürbringer to Dubois, 15 October 1887.

[33] Letters from Fürbringer to Dubois, 26 October 1887 and 2 November 1887.

[34] Bernsen, Journal I, 29 January 1931.

[35] Anon., "Prof. Dubois blikt over zijn leven terug"and Autobiographical Notes.

[36] For Vogt's visit to the Netherlands, and for the stir his lectures created, see Hegeman, "Nederlandse reacties op de evolutieleer", 286–291.

[37] Place, *De descendentieleer*.

[38] De Vries wrote to Dubois on 28 October 1887, shortly before his departure for the Dutch East Indies: "I, too, always remember with pleasure the hours I spent with you, and especially [our] talks about the theory of descent and related subjects."

[39] Lyell described the find in his *Antiquity of Man*, 338–340. K. Martin was the first to express serious objections to the age ascribed to the specimen ("Notiz über den Unterkiefer"). See also Huizinga, "De prehistorische mens in Nederland", 37–38.

[40] Van Bork-Feltkamp, "Anthropological Research", chapter 5.

[41] Only Neanderthal Man attracted a certain amount of attention, albeit indirectly, because of some foreign anthropologists' assertion that the fossils were nothing more than the remains of an archaic Dutchman. A. Sasse, for instance, held that the Neanderthals had lived right up to the present day, and thought that he could produce proof in the form of a skull from North Holland. A. Folmer examined a number of skulls from Friesland, looking for Neanderthal features, but his findings persuaded him to take a sceptical view of Sasse's claim. (Van Bork-Feltkamp, "Anthropological Research", 43, 52).

[42] For a survey of Dutch archaeological research, see: Brongers, "Material".

[43] The Henkeput finds are described by Ubaghs, *Les ateliers*, 16–20, and Cremers, "Excursie Rijckholt". According to Dubois in a letter to A.E. van Giffen, dated 6 August 1925 (*Biologisch-Archaeologisch Instituut*, Groningen), he also found several flint tools and a number of fossil mammal remains in the vicinity of Rijckholt. He donated the tools to the National Museum of Antiquities (*Rijksmuseum van Oudheden*) in Leiden, and the mammal remains to Teylers Museum in Haarlem. The material from the Henkeput, housed in Eijsden castle, was lost during the war; see Schroevers-Kommandeur, "De Henkeput", 119–120.

[44] Autobiographical Notes. The passage quoted ended with "so I thought then." In later years Dubois became convinced that his research into the brain provided direct evidence for "the close links between Man and the animal world." We shall look into this more closely in chapter 5.

[45] As we can see from his Autobiographical Notes, Dubois applied to the Secretary-General of the Colonial Office.

[46] For these reasons, Fürbringer and Weber had most emphatically advised him to drop these plans. Weber had even compared Dubois to the main character in Grimm's fairy tale 'Hans in Luck' (Autobiographical Notes).

[47] Koninklijke Natuurkundige Vereeniging, *Een eeuw natuurwetenschap*, 15.

[48] Dubois, "Over de wenschelijkheid", 1888, 163.

[49] *Idem.*

[50] Fraipont and Lohest, "La race humaine de Néanderthal".

[51] *Ibid.*, 753.

[52] In Virchow's view the bones were those of a modern human. Huxley saw in them only confirmation for his own interpretation in his *Man's Place in Nature.* Most anthropologists regarded the finds solely as proof of the great age of the race. See Erickson, *Origins of Physical Anthropology*, 197–200.

[53] Dubois, "Over de wenschelijkheid", 1888, 155–156, 160–163.

[54] *Ibid.*, 161–162.

[55] *Ibid.*, 162.

[56] Lydekker, "Notices of Siwalik Mammals", 33.

[57] Lydekker, "Siwalik Mammalia", 2.

[58] *Ibid.*, 4.

[59] See for instance Martin, "Ueberreste vorweltlicher Proboscidier"; "Fossile Säugethierreste". For Martin see Brongersma, "Rijksmuseum van Geologie en Mineralogie", 51–57. Brongersma (*ibid.*, 81, note) also mentions a number of obituaries and biographies.

[60] Dubois, "Over de wenschelijkheid", 1888, 149–155.

[61] *Ibid.*, 160.

[62] Lartet, "Note sur les ossements fossiles"; "Nouvelles observations".

[63] Lartet, "Note sur un grand singe fossile".

[64] Gaudry, *Enchaînements du monde animal*, 235–237. Later Gaudry was to withdraw this interpretation: in 1890, in "Le Dryopithèque", he described *Dryopithecus* as a transitional form between monkeys and apes.

[65] Lydekker had expressed his views thus ("Siwalik Mammalia", 3–4): "In those respects in which the Siwalik *Troglodytes* differs from the existing African species it shows in a

still more marked degree the approach to the human type of dentition presented by the latter, and serves, in a small degree, to bind still closer the connection between the Simiidae and the Hominidae. In the inclinations of the two series of cheek-teeth and the relative lateral position of the premolars to the true molars the genus *Hylobates* makes an approach to the human type which is wanting in all the larger existing Simiidae, and it is very noteworthy to find a similar relation obtaining in the Siwalik *Troglodytes*, accompanied by a more humanlike structure of the upper molars."

66 Dubois, "Over de wenschelijkheid", 1888, 160–161.

67 *Ibid.*, 156–159.

68 Two manuscript versions have been preserved, a rough draft and a finished version, both untitled. The rough draft was also used in the compilation of the article Dubois wrote in April 1888 for the *Natuurkundig tijdschrift*. We can conclude from this that Dubois must have given the lecture earlier that year. The quotations from the lecture given in the text are taken from the 36-page final version of the manuscript.

69 For the supposed indications of the existence of Tertiary Man, see Oakley, "The Problem of Man's Antiquity", 117–122; Grayson, *Human Antiquity*, 209–210. For a contemporary survey, see Newton, "Man in the Tertiary Period". Animal bones with cutmarks were discovered in the early 1860s by J. Desnoyers close to Saint-Prest.

70 This is evident from a letter from Dubois to NN, 3 April 1888.

71 Dubois, "Over de wenschelijkheid", 1888, 164–165.

72 *Ibid.*, 165.

73 Details of the exploration of the Lida Adjer Cave can be found in a 12-page unpublished report entitled Provisional Report on Palaeontological Investigations into Caves at Pajakombo (Padang Highlands) (*Voorloopig verslag over palaeontologische nasporingen in grotten bij Pajakombo (Padangsche Bovenlanden)*); this report is in the Dubois Collection.

74 On this committee, see Sirks, *Indisch natuuronderzoek*, 160–161.

75 For this, see Sirks, *ibid.* Dubois' correspondence reveals evidence of the support he was given by the individuals and bodies he named. The support was mainly of a non-material kind, as the committees concerned had no funds to dispense, at that period. Only Kroesen, Governor of Sumatra's West Coast, had given Dubois some actual practical assistance quite early on, by placing several forced labourers at his disposal for the investigation of the Lida Adjer Cave. (Letter Kroesen to Dubois, 8 September 1888.)

76 This can be seen from an undated letter from Dubois to Verbeek. For the report in question, see note 73.

77 It is clear from Dubois' correspondence with those concerned that Verbeek and Groeneveldt in particular played an important part here. Verbeek, whose advice as an expert was sought by the government, gave his full backing to Dubois in his report. Groeneveldt indicated the funds that could be used to finance the investigation.

78 Dubois appeared to have developed malaria.

79 Dubois, "Voorloopig bericht", 1891, 93.

80 Dubois mentioned this experiment in, for example, his report on the Pajakombo exploration (see note 73).

81 Dubois, "Voorloopig bericht", 1891, 93.

82 See note 73.

83 In a letter to Martin, dated October 1889, Dubois mentioned another reason for his original preference for Sumatra: there was no garrison on Java in the neighbourhood of the find-spots. It was also thanks to Verbeek that Java was included in the government assignment as part of the terrain to be investigated. Verbeek had insisted on this in his advice to the government (see note 77). He had also advised Dubois to begin the investigation on Java, since good localities were already known there and the research sites were much more accessible than those on Sumatra (letters from Verbeek to Dubois, 22 November 1888, 8 March 1889 and 9 March 1889).

[84] A report of the discovery can be found in Van Rietschoten, [Uit een schrijven].
[85] Letter from Sluiter to Dubois, 21 December 1888.
[86] Dubois, [Uit een schrijven], 1889.
[87] Dubois revealed the excavation results to the ERI Director by means of monthly, quarterly, and annual reports. The quarterly reports were published anonymously in the *Verslagen van het Mijnwezen*, under the titles "Palaeontologische onderzoekingen ter Westkust van Sumatra" and "Palaeontologische onderzoekingen op Java". The *Verslagen van het Mijnwezen* appeared as supplements to the *Javasche courant*. Dubois' reports on the Wadjak finds are in "Palaeontologische onderzoekingen" (3rd quarter 1890), 1891, 15; "Palaeontologische onderzoekingen" (4th quarter 1890), 1891, 18.
[88] Monthly report for November 1890. In contrast to the quarterly reports, the monthly reports were not published (see note 87). A manuscript version of the monthly reports, in Dubois' hand, is in the Dubois Collection.
[89] Brongersma, "De verzameling van Indische fossielen", 101.
[90] See note 59.
[91] [Dubois], "Palaeontologische onderzoekingen" (1st quarter 1891), 1891, 13.
[92] For further particulars of the excavations, see Brongersma, "De verzameling van Indische fossielen", 100–105; De Vos and Sondaar, "The Importance of the 'Dubois Collection'".
[93] [Dubois], "Palaeontologische onderzoekingen" (3rd quarter 1890), 1891, 13–14.
[94] *Idem.*
[95] *Ibid.* (1st quarter 1891), 1891, 13; *ibid.* (3rd quarter 1891), 1892, 12.
[96] *Ibid.* (3rd quarter 1893), 1894, 15–17.
[97] *Ibid.* (4th quarter 1890), 1891, 15–18.
[98] *Ibid.*, 17.
[99] *Ibid.* (2nd quarter 1892), 1892, 16–17.
[100] *Idem.*
[101] See for instance *ibid.* (2nd quarter 1894), 1894, 12–15.
[102] Letter from Dubois to M. Weber, 29 February 1892. It is obvious from a letter to M. Treub, dated 29 July 1892, that Dubois' original intention was to restrict his work on Java to collecting the fossils and putting them into a provisional kind of order. The final treatment of the material demanded extensive study of the literature and the availability of comparative collections; for this reason it could more easily be carried out in Europe. However, Dubois decided early on to make plans for a first description in the East Indies. One of his reasons for this was the fear that the government would hand over the material to someone else, once the excavations were completed. One likely candidate, in Dubois' view, was Karl Martin, since he had already been involved in working on the Javanese fossil fauna. Dubois also suspected Martin of wanting to cheat him out of working on the fossil finds. His anxiety over government plans were later seen to have been not entirely unjustified, but his distrust of Martin was unfounded. The government did actually sound Martin out about whether he was prepared to take on the description of the fossils, but he turned down the offer in no uncertain terms, with the reasoning that Dubois was the right person for this task (letter from the Minister for the Colonies to Martin, 28 February 1894; Martin's answer of 2 March 1894 (Rijksmuseum van Geologie en Mineralogie, Leiden); there are copies of these letters in the Dubois Collection).
[103] In a letter to R.C. Kroesen on 6 February 1893, Dubois complained of: "Trouble with getting comparative osteological material, still insufficient fossil material, the difficulties (too lightly estimated before) connected with the lack of museums for comparison, and of more abundant literary resources for working on the material in the East Indies... and my state of health, that for a long time has left much to be desired..."
[104] Brongersma, "De verzameling van Indische fossielen", 105; Hooijer, "Fossil Proboscidea", 9–13.

Chapter 3

PITHECANTHROPUS ERECTUS

The discovery

On 24 November 1890, a few months after the beginning of the field investigations on Java, an extraordinary discovery was made at Kedoeng Broeboes, which Dubois described as follows:

> Amidst the remains of typical representatives of the fauna concerned, and in the same layer of sandstone-like andesitic tufa, a human fossil was found, the right side of the chin of a lower jaw with the sockets of the canine tooth and of the first and second premolar.[1]

There could be absolutely no doubt that the fragment was of the same age as the fauna, because: "when it came to me it still had on it a small piece of sandstone-like andesitic tufa, which is the chief sediment among which the bones are found."[2]

Dubois had great difficulty in interpreting the specimen and was unable to reach a final verdict on it. The incomplete state of the jaw was partly to blame for this; it is so fragmentary that discussion over its classification is still possible to this day, making Dubois' uncertainty quite understandable. Yet there were also other factors at work, which had more to do with the framework of Dubois' interpretation than with the nature of the fossil itself.

Dubois was particularly doubtful about whether there was any question of a new human species:

> The chin must have been even less prominent than [it was] in the La Naulette and Schipka [fossils] and because of the very considerable flattening and depression of the inner surface of the lower jaw (caused by the insertion of the *Musculus digastricus*), the concavity being only slightly more marked in Pleistocene jaws from *Europe* than in present-day humans, this fossil [jaw] forms a different and probably lower type than any previously known.[3]

In a provisional survey of the project results achieved in 1890, published in the *Natuurkundig tijdschrift* in 1891, Dubois described the jaw as that of "*Homo, spec. indet.*" Clearly showing his uncertainty he added: "In one respect [i.e., the concavity described above] the difference from the only living species of this genus is so large that it seemed advisable to abandon a closer definition for the time being."[4]

Reading between the lines we find that nineteenth-century notions about fossil and modern human races are clearly visible here. Within this context the interpretation of Wadjak Man had caused Dubois few problems. Now, however, he was confronted with the question of locating the 'lower boundary' of the broad range of forms that could be gathered under the designation *Homo sapiens*. In Dubois' view one feature placed the jaw 'lower' than that of the Neanderthals, until then considered to be the most primitive race of humans. Yet how 'inferior' did a form have to be to fall outside the range of variation in human characteristics?

On 20 December 1890 Dubois mentioned the discovery of the jaw fragment to F.A. Jentink. In his letter he described the fossil human to whom the jaw must have belonged as "not a recent [type]... although nowhere near being a real anthropoid type." It seems that only forms which could be interpreted as unquestionably ape-like had any chance of being located outside the heterogeneous group of *Homo sapiens*. The Kedoeng Broeboes jaw gave Dubois some cause for doubt, but it was not sufficiently ape-like to be allotted its own place. Considering the problems of interpretation it is hardly any wonder that Dubois made no mention of the possible evolutionist significance of the find; there were too many ambiguities.

In August 1891, during the second digging season on Java, Dubois began excavations at Trinil, a locality that was to acquire historical significance. In September 1891 the first remains of a primate, a third molar, emerged from the fossil-rich sediments (see Fig. 8). In his third quarterly report for 1891 Dubois described the specimen as:

> a molar (the third molar of the upper right side) of a chimpansee (*Anthropopithecus*). This genus of anthropoids, occurring only in *West-* and *Central-equatorial Africa* today, lived in British India in the Pliocene and, as we can see from this discovery, during the Pleistocene in *Java*.[5]

He thus considered the molar to have belonged to an ape, seeing a direct link with Lydekker's Siwalik chimpanzee from British India.

The now-famous skullcap (Fig. 6) was unearthed in October by the military engineers:

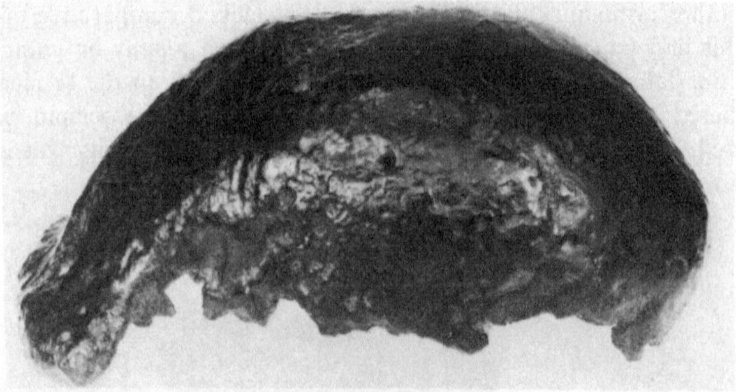

Fig. 6. The Pithecanthropus skullcap.

> Close to the place on the left river bank where the molar was found, a splendid skullcap was dug up that, just as unquestionably as [the] molar, has to be attributed to the genus *Anthropopithecus* (*Troglodytes*).[6]

In Dubois' opinion the skullcap was clearly distinguishable from that of the orang-utan and the gorilla. Although it was not possible to make a direct comparison with the Siwalik chimpanzee, he considered that "a close kinship or identity between both is not impossible."[7] Consequently he adopted Lydekker's genus name *Anthropopithecus* for the Javanese form. (Lydekker had still used the genus name *Troglodytes* in 1886, but changed it to *Anthropopithecus* in 1891.[8])

Despite having designated the find as a 'chimpanzee', Dubois was well aware that he had made an important discovery. The Javanese chimpanzee was no 'ordinary' ape, being substantially more human-like than any known anthropoid. A comparison with the fossil apes known at the time also proved the value of his find, he believed. In 1890 *Dryopithecus*, which had long enjoyed the reputation of being the most Man-like anthropoid, lost this status when it was demoted by Gaudry, on the basis of several new finds, to the level of a transitional form between the Cercopithecidae (Old World monkeys and baboons) and the apes. So, apart from a fossil orang-utan's canine tooth described in 1837 by Falconer and Cautley,[9] Lydekker's Siwalik chimpanzee was "all that we possessed until now in the way of fossil higher anthropoids, and it is clear from this what a gain for science the Javanese fossil is," Dubois wrote.[10] On 30 December 1891 he wrote in a letter to R.C. Kroesen that the being to whom the skullcap had belonged was "truly a new and closer link in the largely buried chain connecting us to the 'lower' mammals."

At first Dubois made no specific comments on the find. Detailed description and placing of the species demanded careful study and comparison with other primate skulls. However, Dubois lacked comparative material, and his first conclusions were consequently based mainly on comparison with the data contained in the literature. Close study of the skullcap was hampered still further because the uppermost part of the occiput was still covered with a thick layer of sediment, while the skull cavity was entirely filled with it (Fig. 7).

Fig. 7. The Pithecanthropus skullcap, still filled with sediment.

Dubois turned to the zoologist Max Weber, asking him to send a chimpanzee skull, but Weber was unable to help him immediately. Nothing was available in either Amsterdam or Leiden, and ultimately Weber had to look abroad. It was October 1892 before he succeeded in procuring a skull in Berlin, and it reached Dubois only on 18 December.[11]

That year, 1891, the Trinil excavations produced no more remains of *Anthropopithecus*. Towards the end of the year, in the monsoon, the rising river waters began to interfere with the work again, and at the beginning of January the investigation had to be suspended.

They were able to resume work at Trinil in mid-May 1892. In August the engineers discovered a third primate fossil, this time an almost completely

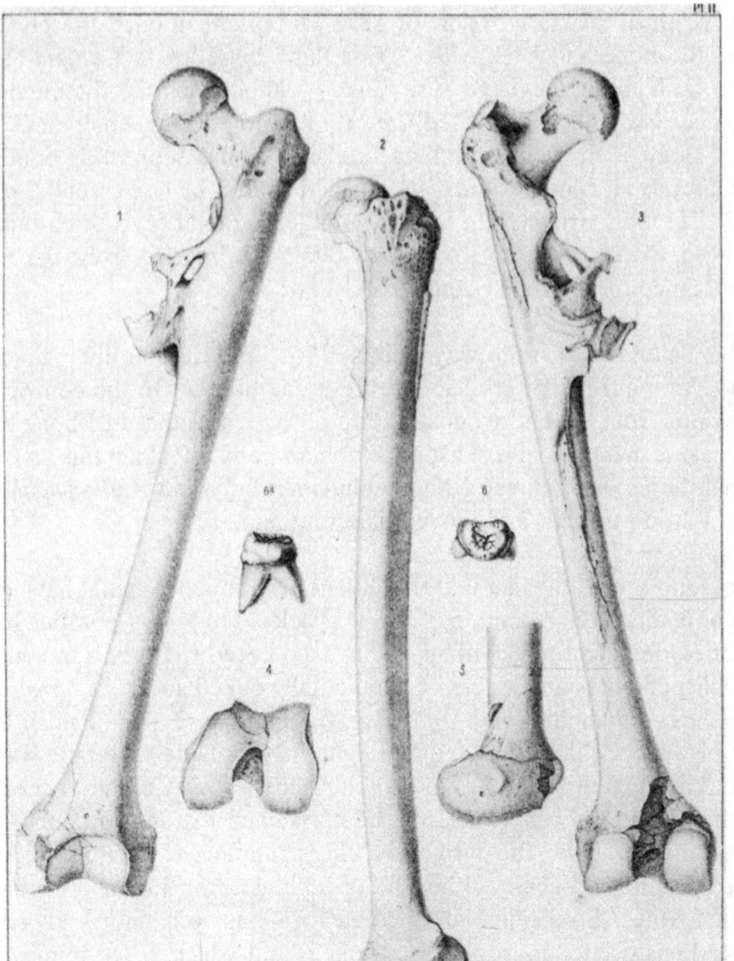

Fig. 8. The Pithecanthropus thigh bone and molar; drawing by Dubois.

preserved left thigh bone (Fig. 8). In fact the specimen was in an even more complete state when it came to light in the sediment than it is today. A mini-drama occurring during the preparation of the bone prevented anyone except the engineers and the forced labourers from ever seeing it in this state. On 7 September 1892 the engineer Kriele had to inform Dubois: "De Winter told me that those small pieces missing from [the thigh bone] were blown away by a strong wind while being glued together on a djati leaf, and could not be found again."

In his third quarterly report of 1892 Dubois mentioned the position in which the bone had been lying: "This thigh bone lay at the same level in which both the other parts were found, yet following the direction of the earlier stream which deposited the tufa material 15 m. upstream." As to morphology, it showed "a striking similarity to this supporting bone of the human body; it is only to be distinguished from it by less essential particulars."[12] Dubois was certain that the skullcap, molar, and thigh bone had belonged to one and the same individual. He inferred from the femur's resemblance to its human counterpart that:

> This being was... in no way equipped to climb trees in the manner of the chimpanzee, the gorilla, and the orang-utan... On the contrary, it is obvious from the entire construction of the femur that this bone fulfilled the same mechanical role as in the human body... Taking this view of the thigh bone, one can say with absolute certainty that *Anthropopithecus* of Java stood upright and moved like a human... [13]

In Dubois' eyes the fact that the Javanese chimpanzee must have walked upright justified the definition of a new species. Up to that point he had left a species identification out of his published quarterly reports, although in his unpublished notes and letters he had actually talked about *Anthropopithecus javanensis* – by analogy with *Anthropopithecus sivalensis* from British India – even before the thigh bone came to light.[14] Now he felt the time was right for making a public announcement about the discovery of a new species: in the third quarterly report of 1892 he christened his find *Anthropopithecus erectus* Eug. Dubois, 'the upright-walking chimpanzee'.[15] Dubois was convinced that he had achieved his aim of finding a transitional form between ape and Man: "So the first intermediate form has been made known in this Lower Pleistocene anthropoid from our island which, in an unmistakable way, links Man more closely with his next of kin among the mammals."[16]

The thigh bone had turned the scales. The skullcap and molar also showed a resemblance to parts of the human skeleton, but this applied to an even greater extent to the thigh bone. For Dubois this was also proof:

> that our ancestors' first step on the road to becoming human must have been the acquisition of the upright position... And consequently the factual evidence is provided that, as some people have already suspected, the East Indies was the cradle of humankind.[17]

On 27 November 1892 Dubois submitted the report embodying these conclusions to his superior, the Director of the ERI department. As we know, however, Dubois' find did not go down in history as *Anthropopithecus erectus*. The official description, published in 1894, was entitled *Pithecan-*

thropus erectus, eine menschenaehnliche Uebergangsform aus Java. When and why did this switch from man-ape to ape-man come about? The published quarterly reports do not present an accurate picture of events. Only in the last report for 1893 did Dubois change to using the name *Pithecanthropus*, without explanation. Before that he had kept to the old name, but it is clear from unpublished manuscripts that he had begun to have other ideas about his chimpanzee as early as December 1892.

Dubois' own version, dating from many years after the event, was that the change of name was inspired by the peculiarities of the femur, which clearly distinguished it from that of the apes but barely at all from that of Man.[18] It was therefore obvious that the Javanese primate must have been much more human-like than Dubois had suspected from looking at the skull's characteristics, so that the title of 'ape-man', Pithecanthropus, seemed more appropriate than 'man-ape'. This was undoubtedly one of Dubois' considerations, but probably not the most important one. There are clear indications that there was a very different situation behind the decision to change the name, namely the discovery of a mistake made in measuring the capacity of the skull, the resulting estimate of the brain volume being consequently far too low. This requires some elucidation.

In the published report for the third quarter of 1892 Dubois estimated the skull volume at approximately 2.4 times that of the African chimpanzee's average capacity.[19] He also mentioned this figure in his official publication on *Pithecanthropus erectus* of 1894. Taking a chimpanzee skull of 410 cc as his starting point he thus estimated the new find at 984 cc, commenting that the real volume might well have been 1,000 cc or more.[20] Yet according to a manuscript version of the third quarterly report for 1892 he had at first quoted the figure of $1^3/_4$ rather than 2.4. The resulting skull capacity would thus be considerably lower, roughly 700 cc. Later – on 4 December 1892 according to a note by Dubois in the margin of the same manuscript – these figures were corrected to the higher values mentioned.

However, Dubois had already submitted a final version of the quarterly report in question on 27 November, which meant that on 4 December he had to inform the ERI Director that:

> In calculating the relative volume of the brains of *Anthropopithecus erectus*, the results being mentioned in the report for the third quarter recently submitted, I made a regrettable mistake. The brains of this transitional form were considerably larger than one would gather from [that report].

Dubois then gave the correct figures, as they would also appear in the published version of the quarterly report, so that his mistake was hidden from the rest of the world.

It is now impossible to decide with absolute certainty what lay behind this incorrect calculation; Dubois' notes and manuscripts give no clear cut answer here. However, we can reconstruct the way in which he may have arrived at his first conclusion. For this the reader is referred to note 21, in which I also explain how Dubois performed the 'correct' calculation of the skull capacity.[21]

Dubois must have discovered his miscalculation between 27 November, when he submitted his quarterly report, and 4 December, when he inserted the correction. Several weeks later, on 28 December, he sent a description of the skullcap, the molar, and the thigh bone to the ERI Director, intended as the first part in a series of publications on his palaeontological excavations on Java. In an accompanying letter he wrote:

> I have, Your Excellency, the honour of offering the first instalment of the description of some of the fossils I have collected. This instalment deals with only one species, Pithecanthropus [sic] erectus.

This slip of the pen is a nice illustration of the fact that the name had been changed only shortly before, and this is proved in a letter from Dubois to Weber on 19 December in which he still talked of *Anthropopithecus erectus*. This means that Dubois only reversed the name several months after the discovery of the thigh bone and, we can assume, only after he had completed most of the description submitted on 28 December.

Dubois' notes and manuscripts from this period provide confirmation of this. In the Dubois Collection there are parts of manuscript versions of Dubois' account of the specimens, in which the fossils are at first described as *Anthropopithecus*, the genus name being later crossed out and replaced with *Pithecanthropus*. Everything suggests that Dubois originally intended to describe the Javanese fossil under the name *Anthropopithecus*, even when the thigh bone had convinced him that he was dealing with a transitional form between ape and Man. It thus seems highly unlikely that it was the conclusions he drew from the thigh bone that prompted Dubois to reverse the name. Rather, it was his discovery that the skull capacity was considerably larger than he had at first thought, so that he suddenly saw the Javanese primate as being much more closely linked with the human.[22]

One result of this interpretation is that the customary view in which Dubois christened his find *Pithecanthropus* in imitation of, or in homage to Haeckel, has to be substantially modified. For a long time Dubois had no intention of changing the name *Anthropopithecus* and when he did he had an 'internal' motive for doing so. Another motive that may also have played a role here, is the possibility of ensuring Haeckel's support by choosing this name above all others, but this was assuredly not the primary cause. If this had been the case, Dubois could have decided on the name *Pithecanthropus* much earlier.

Fig. 9. The Pithecanthropus site at Trinil in 1894.

The excavation work was intensified after the discovery of the thigh bone, but Dubois' hopes that further human-like fossils would come to light proved idle. He only discovered a fossil molar, very worn down and therefore difficult to identify, which after some consideration he attributed to a baboon (*Cynocephalus*).[23]

THE DESCRIPTION

In 1893 Dubois decided to spread word of the discovery of Pithecanthropus officially and widely, in a separate publication. Originally he had wanted to include the description in the proposed series on the whole palaeontological investigation on Java. However, considering the exceptional importance of the find, and because it was becoming increasingly clear to him that his first plan for a series was too high-flown, he abandoned it and confined his attention to the ape-man.

After the excavations of the autumn of 1893 had produced no new human-like remains, Dubois submitted his manuscript in January 1894, and in August 1894 his dissertation *Pithecanthropus erectus, eine menschenaehnliche Uebergangsform*, rolled from the presses.[24] The 39-page work included photographs of the skullcap and thigh bone, and a number of drawings of the third molar and femur made by Dubois himself with the help of a camera lucida. For comparison he provided an illustration of a chimpanzee's and a gibbon's skull.

Dubois began his treatise with a systematic description of *Pithecanthropus erectus* as a species, followed by a concise history of the discovery. He then gave an anatomical description and interpretation of the skullcap, molar, and thigh bone respectively, closing with a section on his general conclusions.

There was only the briefest account of the locality and the circumstances surrounding the discovery. Dubois mentioned that the excavations were carried out close to Trinil on the River Solo, and that the Pithecanthropus fossils were found amid "many remains of the species of Pleistocene mammals and reptiles which have also been found elsewhere in the andesitic tufas of the Kendeng... "[25] The find was made in "the left bank of the Bengawan [the Solo River] approximately 1 m. below the drought level of the river and 12–15 metres below the plain through which this river has carved out its bed... "[26] Dubois observed that the molar and the skullcap had been found at a distance of 1 metre from each other. (In the quarterly reports he had talked only of "some distance.") He gave the figure of 15 metres for the distance between these fossils and the thigh bone, emphasising that all the finds were unearthed from the same layer of sediment and at the same level.

Dubois provided no more information on the circumstances of the find, but in a footnote he said that "It would be foolish (*thöricht*) to doubt [the fact that] the three relics belong together on the basis of this slight distance between their locations."[27] He added that throughout the entire period of excavation, no complete skeleton of any animal species whatsoever had been found, and that skeletal elements that belonged together had been found on several occasions at distances of 20 to 30 metres from each other. Moreover, no skeletal fragment of any organism comparable to the ape-man had been discovered anywhere in the whole excavation area, many thousands of times bigger than the distance between the Pithecanthropus fossils. The only possible exception was the fragment of the lower jaw from Kedoeng Broeboes, which he had at first attributed to a 'low species' of the genus *Homo*: "now [it] appears to me to belong to the form to be described [i.e., Pithecanthropus]."[28] This was all the information he gave on the jaw fragment.

Dubois then described the three fossils, beginning with the skullcap. The skull cavity was still filled with sediment, but the outer surface had been cleaned, meaning that the exact measurements of the cap could be given. Because of the dolichocephalic character of the skull, and the absence of bone crests, Dubois thought that the orang-utan and gorilla could be immediately rejected as comparative material. However, there were obvious similarities to the skulls of the chimpanzee and the gibbon, yet an important difference here – apart from the fossil's bigger measurements – was that the Pithecanthropus skull had a higher vault. In this respect it corresponded to the human. From this Dubois inferred that the facial part of the head must also have been more human-like than it is in the apes.[29]

The form of the eyebrow ridges and the occiput also indicated kinship with the human. Dubois suggested that, where these features were concerned, the ape-man's skull was hardly 'inferior' to those of the Neanderthal and Spy fossils. Yet he immediately added the observation that these skulls were deformed through disease and were consequently of little value as comparative material.[30]

The bottom portion of the back of the cranium bent rather sharply inwards, Dubois noted, causing a torus occipitalis, a 'dent' at the back of the head. This also occurred in gibbons and in young chimpanzees and gorillas, but the bend was much more pronounced in Pithecanthropus than in the apes. Dubois maintained that this was connected with the upright gait.[31]

We have already seen the way in which Dubois determined the skull capacity, and his conclusion that the volume could have been more than 1,000 cc. He also observed that the cranial sutures were completely knitted, showing that the individual must have been an adult and even elderly. It was probably a female.[32]

Dubois thought that the similarities to chimpanzee and gibbon skulls precluded classification within the genus *Homo*, yet on the other hand its capacity set the skull apart from those of the apes. Of itself this did not exclude the possibility that the cranium came from an ape, but the measurements of the femur showed that the brain volume was both relatively and absolutely larger than in the apes. The thigh bone indicated human proportions and weight, while in the gorilla, the largest ape, a much bigger body was combined with a much lower brain volume. Conversely, in *Homo sapiens* a skull capacity of roughly 1,000 cc, which occasionally occurred, always went together with a very small build.[33]

He then turned his attention to the molar. Although this came from an adult it showed few traces of wear. Dubois observed that it displayed symptoms of retrogression, corresponding quantitatively with the retrogression of the human third molar, but qualitatively with the chimpanzee's. From this he inferred that the jaw did not as yet show the characteristically human horseshoe form, yet had a more human appearance than an ape's jaw. Further he drew attention to the fact that the fossil tooth differed from a human molar in some ways, and from a chimpanzee's in others. There could be no conclusive verdict on it, he believed. Incorporation in the genus *Homo* could be ruled out, but it could not be said with any certainty that the molar could not have come from a chimpanzee or gibbon. The possibility that it had belonged to Pithecanthropus remained open, and for Dubois this was the only feasible alternative, given the circumstances of the find.[34]

Finally Dubois discussed the femur. In his view it was human in all its principal features. In its form and measurements, and in all the characteristics relevant to skeletal mechanics, it was indistinguishable from the human thigh bone. The only perceptible differences lay in small details such as the form of the planum popliteum (i.e., the facies poplitea), rounded in the fossil and flattened in the human; the fossil's largely absent angulus medialis and poorly developed linea obliqua; and finally the connecting line between the crista intertrochanterica, straight in the human, concave in the fossil.

> These points suffice, however – as I would emphatically point out – to separate the species in question from Man, who always shows differences in this respect. On these points, minor ones from a mechanical point of view, the femur resembles that of the anthropoids.[35]

Yet in all its other features there were obvious differences from the anthropoids. The thigh bone was much longer and thinner than it was in the anthropoids, with the exception of the gibbons. Anthropoid femurs lacked the torsion along the longitudinal axis, and finally, the joints were differently formed and oriented in anthropoids.[36]

The excrescence seen on the thigh bone had very probably been caused by an external wound, Dubois thought, and he suggested in a footnote that this could have been inflicted by a splinter of wood or by a wooden arrowtip or lance point. This was an implicit declaration that he did not exclude the possibility of the ape-man having used weapons.[37]

After the anatomical description of the three fossil specimens Dubois turned to his general conclusions. First of all he considered the implications of the fossil femur's morphology for bodily organisation. Given the key position occupied by the femur – in Dubois' words the "pillar, girder and siphon" of the body – within the skeletal mechanics, the thigh bone produced important information on the construction of the rest of the body. Dubois posited that the body weight could not have differed substantially from that of the modern human, because the length and thickness of the thigh bone indicated a comparable supporting power. For analogous reasons the body height must have been roughly equal to the modern European's 1.7 metres.[38]

Pithecanthropus must also have walked in the same way as the modern human, and there was consequently no reason to suppose that he had extremely long arms, like the gibbons. Those arms were an adaptation to life in the trees, to swinging from branch to branch. A similar way of moving for Pithecanthropus was a priori improbable, considering this form's much greater weight. Dubois added that on the other hand, the gibbons were the only apes able to walk upright, without using their hands. This proved that there was a close relationship between the length ratio of femur and torso on the one hand, and walking upright on the other. In the gibbon and Man this ratio was approximately the same.[39]

Dubois thought that the foot must also have been human-like given the fully upright movement. The torsion of the femur indicated that the foot was placed flat on the ground, while the chimpanzee and orang-utan, lacking this torsion, walked more on the outer edge of the foot. It also followed that the ape-man no longer needed his front limbs in walking so that arms and hands could have developed towards the human, and could have fulfilled the same function as in Man.[40]

In summing up the fossils' characteristics Dubois concluded that serious objections could be made against the inclusion of *Pithecanthropus erectus* either among the anthropoids or in the genus *Homo*. The ape-man stood between the two:

> an Anthropopithekos has become a Pithekanthropos... *Pithecantropus erectus* is the transitional form which, according to the theory of evolution, must have existed between Man and the anthropoids; he is Man's ancestor.[41]

Dubois now also regarded as proved the proposition already put forward by Lamarck and Darwin that the acquisition of an upright way of walking had been the first step in the process of becoming human. It was only after this that brains and hands had differentiated.[42]

Lastly, Dubois went into the question of the possible line of descent for Pithecanthropus and humankind. He thought that the Pleistocene Neanderthals, who lived in the last interglacial age, were already true humans. This race "stood at a level of morphological development [that was] no lower than that of the modern races of mankind."[43] Once again he suggested that it was impossible to make a direct comparison with the Neanderthals because of their pathological deformation. He added in a footnote that the ape-man's femur was, in its thickness and curvature, even more human than the thigh bones found at Spy, one more proof that these bones, which after all derived from real humans, must be misshapen.[44]

The remains of mammals among which Pithecanthropus had been discovered indicated a dating in the Pliocene or Pleistocene. Dubois' proof for this consisted of a few pieces of faunistic information, the only ones to appear in his entire treatise. He argued that the accompanying fauna did not contain a single typically Miocene species, but that it did include such specialised cattle species as the gaur (*Bibos gaurus*), still living today. Consequently the fauna must be more recent than the late-Miocene or Pliocene Siwalik fauna, in which not a single still-extant species was found. It was probably to be equated with the Narbada fauna of British India, which would give a late-Pliocene or early-Pleistocene dating for the Javanese fauna.[45]

The question now was whether Man could have developed from *Pithecanthropus erectus*. Given the fact that there were already true humans in the last interglacial period, Dubois thought that this development must have been rapid. Towards the end of the Tertiary the living conditions for the mammals, which had differentiated rapidly in this period, became considerably less favourable, resulting in the extinction of innumerable species. A human-like form, with such special adaptations as advanced brain development and upright gait, would have been able to profit from these changed circumstances since these adaptations offered an extra advantage. One had to assume, said Dubois, "that the more rapidly the organs proceed along the road to perfection (differentiation), the greater the advantage that results from it for the organism in the struggle for life."[46] He added in a footnote that this rapid development could have proceeded in jumps: "The results of anthropology point quite definitely to this development of Man per saltum."[47]

Dubois believed that the precursor of Pithecanthropus had to be sought in British India since the Javanese fossil fauna stemmed from the Siwalik fauna, and consequently the most obvious forbear was Lydekker's *Anthropopithecus sivalensis*. Just as with this form, *Pithecanthropus erectus* revealed similarities both to Man and to the chimpanzee and gibbon. The upper jaw of *Anthropopithecus* – the only known relic of this species – already showed obvious traces of differentiation in the direction of the human, and this development must have been continued in the ape-man. Dubois suggested that the African chimpanzee could also have developed from *Anthropopithecus*.[48]

The ancestor of *Anthropopithecus* had probably been a gibbon-like form. Dubois pointed out that gibbons were found exclusively in South-East Asia, while Pithecanthropus showed that human beings had also originated there. Moreover, although the gibbon occupied the 'lowest' position among present-day apes, while the chimpanzee was closest to Man, the gibbon also possessed several features in which it was closer to humankind than to the other apes. Dubois pointed to the morphology of the teeth and the skull, and he also drew attention to the gibbon's limited capacity for walking upright. Apart from its highly specialised arms the gibbon's form was the most generalised among the apes; specialisation for life in the trees had most probably only developed in the fairly recent past. With this information in mind, and in view of the fact that both *Anthropopithecus sivalensis* and *Pithecanthropus erectus* clearly shared certain features in common with the gibbons, it seemed perfectly reasonable to Dubois to assume that the human line had evolved from a Tertiary gibbon-like form. He named this hypothetical ancestor *Prothylobates*, suggesting in a footnote that Lartet's *Pliopithecus* might possibly be regarded as this primaeval form. Thus the human family tree acquired the following outline: *Prothylobates – Anthropopithecus sivalensis – Pithecanthropus erectus – Homo sapiens*.[49]

Of one thing Dubois was certain: the proof had been found for the proposition that humankind had evolved from a line of apes. The ape-man was the transitional form that had bridged the gulf between man-ape and Man.

Dubois' treatise on *Pithecanthropus erectus* can be called a milestone in the development of palaeoanthropology; I shall explain the reasons for this in the next chapter. Here, I shall address only the treatise itself, giving an analysis of Dubois' work, and casting some light on the way in which he arrived at his interpretation of the fossils.

As we saw, Dubois was quickly forced to abandon his plan to publish the results of his Sumatran and Javanese excavations while he was still in the East Indies; this obviously required more literature and comparative material than lay within his reach on Java. The lack of reference material also presented a handicap in describing the Pithecanthropus fossils. Dubois had with great difficulty managed to secure a chimpanzee skull through Weber's good offices. His treatise shows that he had also acquired three gibbon skulls, but for the rest, as far as the anthropoids were concerned he had to make do with the existing literature and the illustrations it contained. He had the same difficulty with human bones; he did have the use of some human skeletal material, but he had no access to comprehensive collections of human comparative material, absolutely indispensable for producing a well documented verdict. Neither did he have any Neanderthal fossils, or at all events casts of them, close at hand. True, he did not regard the latter as a serious drawback since he considered the Neanderthal and Spy remains to be deformed, but, on the other hand, if he could have inspected the bones for himself he might have changed his mind. Finally, Dubois knew the fossil anthropoids only from the literature.

He was better supplied where literature was concerned. As we can see from his correspondence, he was able to obtain the books and most of the journals he needed for his description from the libraries of the Dutch Indies Mining Company and the *Natuurkundige Vereeniging*. The bibliographical references in his treatise show that he had access to a reasonably good collection of anatomical and osteological handbooks on the apes and humankind.

Nevertheless, all in all Dubois' initial position was not at all a favourable one, and we must not lose sight of this fact in what follows. In the preface to his treatise Dubois acknowledged that an exhaustive analysis of the fossils was beyond his reach in the East Indies. Yet, as he said himself, a provisional account could be delayed no longer given the exceptional importance of the finds. It is also probable that the government was pressing him to publish while he was still on Java. This would not have been surprising, since nothing had come of the somewhat unrealistic promises to publish that Dubois had been making to the government since 1892. It may have been a matter of urgent necessity to publish at least the description of the Pithecanthropus fossils if he wanted to avoid losing all credibility in governmental eyes.

Dubois' treatise on *Pithecanthropus erectus* is unmistakably the work of an anatomist. So much is clear merely from his description of the circumstances in which the finds were made. In fact Dubois was content with providing an indication of the geographical location of the find-spot and the statement that the molar, skullcap, and thigh bone were found in the same layer, at a distance of one and fifteen metres respectively. More detailed information on the position of the fossils, for instance in the form of a profile drawing, was lacking. Neither did he provide anything more than casual information on the fauna among which the fossils had been found. His recording of the circumstances surrounding the discoveries on Java was superficial, to say the least, when compared with De Puydt's and Lohest's extremely precise documentation of the Spy finds. The description of the Neanderthal fossils from Spy included a complete specification of the accompanying fauna and a profile drawing of the excavation site in which the position of the bones was shown. There was even an attached eye-witness declaration from people who had been taken to the find-spot expressly to confirm the location of the fossils.[50]

Dubois' method of excavating rendered such precision impossible from the very beginning. He did not perform the actual digging himself and was only occasionally present at the locality. At this stage of the excavations, the work was being carried out without any mapping of the site, with quadrants for instance. Dubois was thus entirely dependent on the engineers Kriele and De Winter for any information they could give him on the fossils' position, with all the attendant inaccuracies likely to arise. For example, because of this way of working there has never been any real certainty over the distance between the thigh bone and skullcap. At first Dubois gave a figure of ten metres for this, subsequently changing it to twelve metres and finally to fifteen. Years later he was to have fresh doubts and to return to his estimate of twelve metres.[51]

Even leaving these uncertainties aside, it is clear from Dubois' quarterly reports alone that he possessed far more information about the locality than he brought into his treatise. He could have devoted more attention to this aspect, but his priorities evidently lay elsewhere. As an anatomist he concentrated on the morphology of the fossils. It was decidedly unwise, in view of the very brief description of the site, to venture the remark that it was *thöricht* to doubt that the fossils belonged together. No wonder that many of his critics immediately took up the gauntlet he had thrown down under their noses.

Dubois followed the tradition of the German school of morphology in describing the fossils. The principal aim of the work was to determine the phylogenetic position of Pithecanthropus and to reconstruct the human family tree, as far as possible. For this Dubois relied on morphological analysis and comparison with related groups of primates.

Like most phylogenetic-morphological studies from the second half of the nineteenth century, Dubois' work had an unmistakable typological bias. Not that his point of departure was a limited number of 'ideal' types providing the model for the entire animal kingdom – if asked, he would undoubtedly have denied the existence of such types – but this outlook was implicit and probably even unconscious. This requires some elucidation.

Coleman has shown that the roots of phylogenetic morphology, which flourished after 1859, were firmly anchored in the typological morphology of the pre-Darwinian era.[52] The metamorphosis taking place within morphology under the influence of Darwin's evolutionary theory amounted essentially to the replacement of formal relationships between types by genealogical relationships. Where structural similarities between organisms had originally been interpreted as indications of typological correspondences, they were now regarded as a sign of common descent. Where the descriptive part of the investigation was concerned there was relatively little change, and although the interpretative context was changed, typological thinking was decidedly still there in the background.

Palaeontological research, especially, bore the heavy imprint of the typological inheritance.[53] Given that attention was directed predominantly towards reconstructing lines of descent, the nature of palaeontological research material did much to foster a typological approach. Complete series of fossils, from successive geological periods and in successive stages of development, were very scarce, and the palaeontologist usually had to make do with fragmentary material of completely different ages. For a long time there was often no answer to the questions of how characteristics had developed over the course of time, of which selective pressures had been responsible for this, or of the extent to which a characteristic could be employed to indicate phylogenetic kinship. Studies such as those by Vladimir Kovalevskii, who was able to verify how, and under what forces of selection the horse's hoof had evolved over the ages, were only possible on a limited scale.[54] Nevertheless the inadequacy of palaeontological material must not be held solely responsible for the typological working method within nineteenth- and early twentieth-century palaeontology. The chief cause was the morphological tradition combining a preoccupation with reconstruction of genealogical trees with thinking in types.[55]

There are clear traces of this thinking-in-types in the account of *Pithecanthropus erectus*. In his investigations Dubois was confronted by the limitations of palaeontological material, and this will have further intensified his inclination towards a typological approach. For instance, he spoke of *Affentypus* and *Menschen-typus*, and employed these concepts as though they were precisely and conclusively defined entities. The difference in approach from typological morphology is evident only in the interpretation, when 'apelikeness' and 'humanlike-ness' were presented as proof, not of typological similarity, but of phylogenetic kinship.

It is also illuminating that Dubois barely restricted himself in the phylogenetic interpretation of form characteristics, each one of which was given a label, so to speak, indicating whether it showed kinship with ape or human. The extent to which the form characteristics in question could provide a valid indication of phylogenetic kinship, and the extent to which there were other possible causes for similarities of form, were seldom if ever discussed. One of the consequences of this was, for example, Dubois' idea that there was a close relationship between Man and the gibbons, and Dubois was certainly not the only one to hold this view.[56] Even though he was well aware that the gibbons are specialised climbers and that in some respects they resemble the lower apes more than the great apes, he nevertheless perceived an indication of phylogenetic kinship with *Homo sapiens* in their (limited) capacity for walking on two legs.

Dubois' systematic classification of Pithecanthropus also shows his typological approach. In his view the ape-man was a predecessor of Man and belonged to an advanced stage of the journey towards full humanity, yet such beings could not be classified as Hominidae because of a number of still 'typically' ape-like features. Neither was there any justification for the ape-man's inclusion in the Simiidae. The skull's features were in fact consonant with such a classification, but the 'typical' human femur militated against it. Dubois believed that classification either with the apes or with humans was inconsistent, and that Pithecanthropus thus represented an independent family, a separate type, so to speak. In Dubois' own words:

> Although already quite advanced in the development of the human type (*Typus*), this Pleistocene form had not yet reached it. – It occupied a position between this one [i.e., the human type] and the type of the great apes... Considering all these circumstances I feel compelled to place the species in a new genus – *Pithecanthropus* – but also in a new family – Pithecanthropidae – between the Hominidae and the Simiidae... [57]

We saw in the first chapter that fossil human remains were classified with the modern human until the 1880s, and at most were ascribed to a separate race. Dubois adhered to this example where Neanderthal Man and Wadjak Man were concerned. In the case of the Kedoeng Broeboes jaw, however, we have seen him doubtful. Could this jaw be assigned to a 'low' race of humans, or did it cross the boundaries of variation among the human 'type'? Here Dubois found himself confronted with the question of where the border actually lay: how much did a fossil have to deviate from the 'type' to be allotted a separate status? This raises the question of why Pithecanthropus did convince Dubois, whereas the fossils just mentioned did not. To what did the ape-man owe his special position, in his view?

We can gather from his first comments on the fossils directly after the discovery, and also from his treatise, that the ape-like character of the skullcap (ape-like in Dubois' eyes, at all events) convinced him that he was not dealing with human bones. He saw the skullcap as undoubtedly of the 'ape type', hence the classification in the genus *Anthropopithecus*. Even after the discovery of the femur, which made him absolutely certain that he had found a transitional form between ape and human, he continued at first to maintain that it was a chimpanzee. His correction of the figure for the skull capacity made him decide to reverse the name, but this in no way made the skullcap a human one. Its ape-like morphology remained the argument for denying it human status.

Dubois' treatise shows clear signs of this train of thought. It is striking that in his anatomical description of the skullcap he compared it almost exclusively with ape skulls, and he continually used chimpanzee and gibbon skulls as reference material. He gave their measurements in his tables and then contrasted them with the Pithecanthropus skull. Dubois did point out resemblances between the skullcap and the human skull, but any direct comparison with an actual human cranium was restricted to a figure comparing profile sketches of the fossil and a human skull. Finally, Dubois took a chimpanzee and a gibbon skull as his standard for judging the relationship between skull measurements and volume in his calculation of the skull capacity, using no human example in his computations. The conclusion he drew from his analysis of the fossil cranium speaks volumes: "From the investigation of the skull alone it does not follow, however, that the fossil form does not belong to the genus *Anthropopithecus* or *Hylobates*."[58]

Thus Dubois thought of his missing link as an ape-like form which because of its development in the direction of Man could be regarded as a transitional form, rather than as a human-like form still at too low a stage of development to be called *Homo*. In the chronology of the discovery it was the femur which assured Dubois that he had discovered a transitional form, but conceptually speaking the ape-like nature of the skullcap was crucial. In historical terms this means that Dubois' conception of *Pithecanthropus erectus* as an intermediate form was also partly due to – or can be attributed to – the continued effect of typological thinking within phylogenetic morphology. At all events it was precisely the typological element in Dubois' approach that induced him to characterise the skull as 'belonging to the ape type'.

If one were to take Dubois' treatise out of its historical context, it might encourage the suspicion that he was exaggerating the ape-like nature of the skull, but if this had indeed been the case it would only have confirmed the assumption that an 'ape-like nature' constituted an essential ingredient of Dubois' conception of a missing link. Yet the facts give the lie to the suspicion. Dubois' account clearly reflected his methodological approach and in this sense was in no way contrived. Moreover he had pronounced his verdict on the skullcap before the discovery of the femur. He immediately described it as ape-like and in fact he always adhered to this opinion.

Dubois did perhaps exaggerate a little in disqualifying Neanderthal Man as comparative material. It is true that he viewed this form as entirely human and thus had no more need for a thorough comparison with the fossil than in the case of the modern human. Yet it still looks as though he made things easier for himself by portraying the Neanderthals as pathological, thus avoiding the necessity for comparison. As we saw in the last chapter, he was much less certain on this point than he appeared to be in his publication. On the other hand, for Dubois the pathological character of the Neanderthal fossils may have been confirmed by the more human appearance of the Pithecanthropus femur.[59] But whether pathological or not, Neanderthal Man was set aside in the first place because of his humanity. This also squares with the fact that, as far as we can tell, Dubois made absolutely no attempt to obtain casts of the Neanderthal fossils, while he did make every effort to acquire a chimpanzee skull.

In his treatise Dubois did not discuss his Wadjak finds either. He also regarded these as human, and therefore the same explanation applies as for the Neanderthals. He did make a very glancing reference to the jaw fragment from Kedoeng Broeboes which he had earlier attributed to the genus *Homo*, now suggesting that this fossil might also have belonged to Pithecanthropus. It is fairly easy to fathom the reason for this change of opinion. It 'must have belonged' to Pithecanthropus because Dubois believed that the Kedoeng Broeboes sediment layers were as old as those at Trinil. The existence of a *Homo* fossil in layers of the same age as those in which the ape-man had been found, would have considerably weakened his argumentation that *Pithecanthropus erectus* was a transitional form between ape and Man. It was obvious, however, that his doubts about the jaw fragment had not dissipated, since he omitted any description or illustration; we can imagine that he was anxious not to present his critics with an easy target.

One also suspects that tactical considerations lay behind Dubois' remark that humanity could have developed from Pithecanthropus 'per saltum', by abrupt leaps forward. His reason for referring to this mechanism was probably that it could provide an escape route should the period of time between the appearance of Pithecanthropus and that of humankind be regarded as too short for gradual evolution from one form to the other. At that time the duration of the Pleistocene was estimated at a considerably lower figure than it is today.[60] Moreover, no one knew for certain at what point in the Pleistocene the first modern humans had come into being.[61] It is obvious from Dubois' observations that his point of departure was the idea that humanisation must have proceeded relatively rapidly after the Pithecanthropus phase.

It is not entirely clear where Dubois obtained the idea of abrupt change,[62] but the idea was certainly not a new one. Given the deficiencies of palaeontological material it is hardly surprising that, in palaeontological circles especially, there was a sceptical view of Darwin's gradualist and selectionist conception of the evolutionary process. As Rainger has demonstrated with the help of many examples, alternative principles such as saltationism, orthogenesis, and Lamarckism prevailed until the twentieth century, especially among palaeontologists.[63] Even Huxley, who is often depicted as the prototypal Darwinist, was inclined to accept a saltatory model; there is also a distinct typological element in his palaeontological work.[64] Dubois' views fit into this picture. Even though he was a dedicated follower of Darwin in those years, as we can see from his Padang lecture for instance, he evidently did not exclude the possibility of the parallel action of other mechanisms besides that of Darwinian selection.

Earlier I touched briefly on the idea that Haeckel's influence on Dubois may well have been more qualified than has usually been believed up to now. The chief, if not the only, reason for linking Dubois' name with Haeckel's has always been the fact that Dubois adopted the genus name *Pithecanthropus*, first suggested by Haeckel, for his Trinil find. Yet only very limited significance can be attached to this fact. Authors such as Walther, for instance, who sent Dubois in search of the missing link, "filled with enthusiasm for Haeckel's world of thought," or such as Von Koenigswald, who believed Dubois to have formulated his ideas on Pithecanthropus under the influence of "Haeckel's theoretical views," give a distorted picture of his views and furthermore take no account of the matter in which Dubois contrasts with Haeckel.[65]

There is no denying that Dubois was influenced by Haeckel, but we are more concerned here with 'influence' in the general sense, as the majority of the biological researchers in Dubois' time experienced it, to a greater or lesser extent. All self-respecting biologists read Haeckel and encountered his ideas, and many of them became acquainted in this way with the theory of evolution and with phylogenetic morphology. Haeckel's influence was also to a large extent responsible for the monopoly enjoyed by morphological research in the second half of the nineteenth century. He aroused many scientists' interest in this research field and many of them indeed devoted their time to phylogenetic studies.[66]

Haeckel's works made a deep impression on Dubois as well, as we can see in the following quotation from a letter to Haeckel on 24 December 1895:

> I should like to tell you how happy I am to be able [by the discovery of Pithecanthropus] to express my gratitude for the influence which you, especially by your "Schöpfungsgeschichte" have exerted on the whole course of my life.[67]

Dubois belonged to the school of morphology founded by Haeckel and Gegenbaur. He was no more averse than Haeckel to speculative reconstructions of possible lines of descent or to the insertion of hypothetical intermediate forms, as we can see clearly from Dubois' phylogenetic tree for Pithecanthropus. For example he proposed a hypothetical form, *Prothylobates*, with gibbon-like attributes, as the ape-man's far-distant ancestor.

However, there are obvious differences to be seen. Dubois never subscribed to Haeckel's monistic-materialist philosophy, believing that the scientific investigator should refrain from philosophical considerations. Like religious belief and science, philosophy and science should be kept apart. Haeckel with his monism had strayed outside the area of science and Dubois saw no advantage in that direction.[68] "Philosophy, from Aristotle on, has not taken the world any further," he once said.[69] He would have had little objection to Haeckel's penchant for audacious theories and daring hypotheses, but he did not follow him into his philosophical speculations; in this sense Dubois was no Haeckelian.

It is evident that Dubois' ideas on 'philosophy' were not very subtle. It is precisely because he did not want to meddle in philosophy that nowhere in his work can we find a declaration of precisely what he considered it to be. The impression one gains is that he considered his own work to be entirely free of it. Nevertheless there can be no doubt that – at all events in the period under discussion here – Dubois took a considerably more empirical stance than Haeckel. He certainly did not fight shy of theories and hypotheses, but he did not want to lose sight of the link with the empirical.

Dubois' attitude towards the question of human descent illustrates this. He did not confine his attention to the formulation of hypotheses on the forms connecting ape and Man that might have existed, and he was not satisfied merely with the indirect clues to these provided by comparative anatomy and embryology. His objective was to unearth the transitional forms themselves, and to illustrate the process of humanisation on the basis of direct fossil evidence. This is why he opted for a palaeontological approach. The contrast with Haeckel on this point is evident. Haeckel regarded human descent from the apes as a proven fact and in his opinion the process was sufficiently documented through comparative anatomy and embryology, rendering unnecessary the search for fossil transitional forms. From his point of view the results of such ventures were of limited value, certainly when compared with the results obtainable with much less effort through anatomical and embryological study. In contrast Dubois yearned for concrete evidence, and in this respect his position agreed more with Gegenbaur's, who likewise did not go so far as to claim that evolution and descent were facts requiring no further proof.

It was an empirical datum – the skull capacity that turned out to be higher than originally expected – which decided Dubois to reverse the name *Anthro-*

popithecus. There was no question of a direct influence from Haeckel's 'world of thought'. In the second place it does seem likely that opportunist considerations also played a part. In giving shape to Haeckel's brain child Dubois was probably hoping to secure his support in defending the discovery. As we shall see, he achieved the desired effect; Haeckel was far from indifferent to the tribute.

Dubois' choice of a palaeontological approach to the problem of humanisation is undoubtedly his chief scientific merit. Not only Haeckel but the majority of morphologists and anthropologists in the second half of the nineteenth century saw no benefit in such an approach, or else they found it too risky. Naturally Dubois was only able to make his contribution because he had quite extraordinary luck in achieving his aims, but this in no way detracts from his merit. Whether or not it was in defiance of all expectations, Dubois did find Pithecanthropus and in so doing was the first to try to reconstruct the process of Man's development on the basis of fossil remains of a human-like predecessor.

Notes

1 [Dubois], "Palaeontologische onderzoekingen op Java" (4th quarter 1890), 1891, 14. L.D. Brongersma has drawn my attention to the fact that there is room for doubt about whether Kedoeng Broeboes was actually the place where the jaw fragment was found. In his diary for 1890, housed in the Dubois Collection, Dubois mentions Kedoeng Loemboe rather than Kedoeng Broeboes as the find-spot. However, to avoid confusion I shall follow the usage in the literature and continue to speak of Kedoeng Broeboes.

2 Unpublished report for November 1890.

3 [Dubois], "Palaeontologische onderzoekingen" (4th quarter 1890), 1891, 15.

4 [Dubois], "Voorloopig Bericht", 1891, 94–95.

5 [Dubois], "Palaeontologische onderzoekingen" (3rd quarter 1891), 1892, 13–14.

6 *Ibid.* (4th quarter 1891), 1892, 13.

7 *Ibid.*, 13–14.

8 Flower and Lydekker, *Introduction*, 738.

9 See Lydekker, "Siwalik Mammalia", 4.

10 [Dubois], "Palaeontologische onderzoekingen" (4th quarter 1891), 1892, 14–15.

11 Letters from M. Weber to Dubois, 4 April 1892 and 21 October 1892. Letter from Dubois to Weber, 19 December 1892.

12 [Dubois], "Palaeontologische onderzoekingen" (3rd quarter 1892), 1893, 10.

13 *Ibid.*, 12–13.

14 For example, in a letter to the Director of the ERI department dated 27 August 1892.

15 [Dubois], "Palaeontologische onderzoekingen" (3rd quarter 1892), 1893, 11.

16 *Idem.*

17 *Ibid.*, 11, 14.

18 Dubois wrote to G. Heberer on 24 September 1934: "Only the discovery of the femur made me gradually recognise the Trinil primate's very close likeness to the human, and its dissimilarity to the chimpanzee, as a consequence of which I reversed the generic name given to the fossil species from Anthropopithecus to Pithecanthropus." Quoted in Walther and Heberer, *Im Banne Ernst Haeckels*, 149.

[19] [Dubois], "Palaeontologische onderzoekingen" (3rd quarter 1892), 1893, 11.

[20] Dubois, *Pithecanthropus erectus*, 1894, 10–11.

[21] First of all the calculation as Dubois made it in his 1894 book on Pithecanthropus. Since the skull cavity was still full of sediment, it was impossible to determine the capacity of the skullcap in the traditional way, for instance by filling it with peas, water, or leadshot. To obtain a rough estimate, Dubois first determined the relationship between the measurements of the skullcap and those of a chimpanzee skull depicted by Theodor Bischoff (*Ueber die Verschiedenheit in der Schädelbildung*, plate V). In his view, where overall form was concerned the cap could best be compared with this chimpanzee's. The ratios of the length and the width of the fossil skullcap to those of Bischoff's chimpanzee were both approximately 1.33 : 1. So the contents of both skulls, if one assumed complete uniformity, would bear the ratio $1.33^3 : 1^3$ or 2.35 : 1. However, the fossil skull had a higher vault than that of the chimpanzee, and consequently there could be no question of complete uniformity; the Pithecanthropus skull must have had a larger capacity. Taking as an example the comparison between a male and a female chimpanzee skull, Dubois showed that the effect of a higher vaulting could be very considerable, and could result in an extra volume amounting to more than one-seventh of the capacity of the least vaulted skull. Because of this the ratio between Pithecanthropus and the chimpanzee could be fixed at roughly 2.4 : 1. Given that the capacity of the chimpanzee skull with which it had been compared was 410 cc – a value that tallied with the average capacity among these animals – then the Pithecanthropus skull volume must have amounted to 984 cc, at least. According to Dubois, the higher skull vault of the latter could also have produced a volume of 1,000 cc or more. On the basis of a gibbon skull conforming equally well in form to the Pithecanthropus skull, Dubois carried out an analogous computation, producing a result of 991 cc, which was gratifyingly close to the first result. How then, did it come about that in his original calculation Dubois reached a volume ratio of 1¾ : 1 between Pithecanthropus and chimpanzee, meaning a skull capacity of approximately 700 cc for Pithecanthropus? Although we can no longer discover with any certainty how the fault arose in the calculation, a likely answer is that Dubois misinterpreted the length of the chimpanzee skull as given by Bischoff. This length was 137 mm, Bischoff meaning the total length (*Ueber die Verschiedenheit in der Schädelbildung*, 54). Dubois probably thought, erroneously, that this meant the length of the skull cavity, in other words the internal length. Then he must also have used the internal length in measuring the fossil skull, which came to 155 mm. Calculated this way the ratio of the linear measurements of Pithecanthropus to those of the chimp was 1.13 : 1. The skull-volume ratio would then be roughly 1½ : 1, and, compensating for the higher vault of the fossil, this brings us to a capacity of approximately 700 cc for Pithecanthropus.

[22] On 18 December 1892 the long-expected chimpanzee skull sent by Max Weber reached Dubois, allowing him to compare the fossil for the first time with its putative 'next of kin'. However, this comparison cannot have been the *experimentum crucis* for the change of name. In 1894 Dubois still held that the form of the skull tallied with that of the chimpanzee's. The important difference was in the skull capacity.

[23] [Dubois], "Palaeontologische onderzoekingen" (4th quarter 1892), 1893, 11.

[24] Letters from Dubois to the ERI Director, 18 January 1894, and to Th.F.A. Delprat, 10 September 1894.

[25] Dubois, *Pithecanthropus erectus*, 1894, 1.

[26] *Idem.*

[27] *Ibid.*, 2.

[28] *Idem.*

[29] *Ibid.*, 2–6.

[30] *Ibid.*, 6.

[31] *Ibid.*, 7.

[32] *Ibid.*, 10–13. See also note 21.
[33] *Idem.*
[34] *Ibid.*, 13–16.
[35] *Ibid.*, 16–23, especially 23.
[36] *Ibid.*, 23–26.
[37] *Ibid.*, 18.
[38] *Ibid.*, 26–27.
[39] *Ibid.*, 28.
[40] *Ibid.*, 29–31.
[41] *Ibid.*, 31.
[42] *Ibid.*, 32–33.
[43] *Ibid.*, 34.
[44] *Ibid.*, 34–35.
[45] *Ibid.*, 35–36.
[46] *Ibid.*, 36.
[47] *Ibid.*, 37. Dubois referred here to a lecture which D.G. Brinton was supposed to have given at the American Asssociation for the Advancement of Science in 1893. There must have been a mistake here, since Brinton did not give a lecture of this kind. I have been unable to discover what Dubois did actually mean.
[48] *Ibid.*, 37–39.
[49] *Idem.*
[50] De Puydt and Lohest, *l'Homme contemporain du mammouth a Spy*; Fraipont and Lohest, "La race humaine de Néanderthal".
[51] In his report to the ERI Director for August 1892 Dubois mentioned a distance of ten metres. He changed this to twelve metres, probably in response to a letter from Kriele, on 7 September 1892. Yet he mentioned a figure of fifteen metres in the report for September 1892 and in his official publication on Pithecanthropus. In a much later publication ("Die phylogenetische Grosshirnzunahme", 1930, 268) he once again brought up the figure of twelve metres.
[52] Coleman, "Morphology between Type Concept and Descent Theory". For an exposition of the various ways in which the type-concept was employed in zoology, see Farber, "The Type-Concept in Zoology".
[53] For the effect of pre-Darwinian conceptions on palaeontology after 1860, see Rainger, *Understanding the Fossil Past*, Introduction, *passim*.
[54] For Kovalevskii's palaeontological work, see Todes, "V.O. Kovalevskii"; Rainger, *Understanding the Fossil Past*, 110–128. For several other examples see Russell, *Form and Function*, 359–364.
[55] See note 53.
[56] In the next chapter we will be discussing several other authors who believed in this.
[57] Dubois, *Pithecanthropus erectus*, 1894, 31.
[58] Dubois, *ibid.*, 12.
[59] For example, Dubois observed (*ibid.*, 35): "That the bent position (*houding*) of the knee of Spy man can only be ascribed to pathological malformation, now also follows from the fact, that *P.erectus* did not possess this 'attitude plus pithécoide qu'humaine'... it was more man-like in this form than in the man from Spy!"
[60] In 1863 Lyell's estimate of the duration of the Pleistocene was 800,000 years; Dana's in 1874 was 720,000; Walcott's in 1893 400,000; Heim's in 1894 100,000; and Sollas' in 1900 400,000. See Osborn, *Men of the Old Stone Age*, 22.
[61] For further particulars see Oakley, "The Problem of Man's Antiquity", 95–104, 123–131. According to Dubois the oldest evidence for the presence of humans dated from the last inter-glacial period (*Pithecanthropus erectus*, 1894, 35). He was probably basing his state-

ment on Ranke's *Der Mensch* (II, 418–420), a work he also mentioned in his treatise, albeit in a different connection.

62 In an interview in 1938 (Anon., "Prof. Dubois blikt over zijn leven terug") Dubois observed that, during his time in the East Indies, Hugo de Vries kept him informed on his "ideas about the mutation theory." This suggests that Dubois could have derived his ideas about 'evolution per saltum' partly from De Vries. The latter published his mutation theory only after the turn of the century, but the *Oenothera* investigations had already been begun in the later 1880's. In 1888 De Vries believed that he had discovered the first macromutations ('monstrosities' as they were then called) in his cultures (see De Veer, *Hugo de Vries*, 147–171). Nothing more can be said with any certainty about Dubois' contacts with De Vries during his stay in the East Indies. Only one letter has survived from all their correspondence in the relevant period, a letter from Dubois on 9 November 1894, thanking De Vries for an offprint. (A transcript of this letter is held in the Dubois Collection. No letters from the period in question are available in the Hugo de Vries laboratory.) We can however gather from Dubois' letter that he and De Vries regularly sent each other their publications.

63 Rainger, *Understanding the Fossil Past, passim.* See also Coleman, *Biology in the Nineteenth Century*, 80. For more information on the alternative mechanisms that were proposed to explain the process of evolution, see Bowler, *The Eclipse of Darwinism.*

64 Di Gregorio, "The Dinosaur Connection"; *Huxley's Place in Natural Science*, 65–68. See also Bartholomew, "Huxley's Defence of Darwin".

65 Walther and Heberer, *Im Banne Ernst Haeckels*, 115; Von Koenigswald, "De eerste 'aapmensen'", 243; Weinert obviously thought that Dubois had come straight from under Haeckel's wing, since he wrote (*Menschen der Vorzeit*, 33): "Around 1880 the Dutchman Dr. Eugen Dubois was an assistant at the Anatomical Institute at Jena and thus became acquainted with Haeckel's theories of phylogenetic trees."

66 For Haeckel's role as promoter of phylogenetic-morphological research, see for example Schmidt, *Was wir Ernst Haeckel verdanken* and Uschmann, *Geschichte der Zoologie.*

67 Quoted in Walther and Heberer, *Im Banne Ernst Haeckels*, 149.

68 Bernsen wrote for instance (Dagboek I, 26 January 1931): "Dubois thinks it is wrong to mingle philosophy with science, neither should a philosopher dress up as a scientist nor a scientist as a philosopher."

69 Bernsen, Dagboek III, 18 January 1932.

Chapter 4

THE DEBATE

Introduction

Dubois' treatise on *Pithecanthropus erectus* did not go unnoticed. By the end of the nineteenth century nearly eighty articles and books had been published in which the fossils were discussed, and this applies only to reactions from the scientific world. As is well known, opinions on the discovery varied widely, virtually every conceivable view being defended by one person or another. Many commentators were doubtful about whether or not the skull-cap and thigh bone belonged together. Some attributed the skull and molar to an ape, others to a human, some to an ape-man. The same applied to the thigh bone, although according to prevailing opinion it was human in origin.

In the existing literature on Dubois, attention to the debate is usually confined to a catalogue of the different interpretations of the fossils. In this chapter I shall try to analyse the reception of the find more closely, elucidating the background to the responses, the historical context of the discussion, and the way in which the debate developed over the course of time. Finally I shall consider the significance of Dubois' discovery for the subsequent development of palaeoanthropology.

For this purpose a somewhat detailed sketch of the reactions to Dubois' work is required, and I shall begin with this. I am not aiming here at a comprehensive account of the reactions, but at a representative picture. Apart from this I shall confine my attention to the way in which the discovery was received in scientific circles. We are thus dealing with researchers who gave a professional opinion on the fossils and published them in scientific journals and books. Popular articles and exclusively reference works will not be dealt with, although the boundaries can by no means always be sharply drawn. The criterium that will weigh most heavily here is the presence of a personal, well-argued opinion on the fossils. A final restriction will be that the post-1900 discussion of Pithecanthropus will not be treated exhaustively. In view of our objective it will be sufficient if we provide a detailed discussion of publications appearing up to the turn of the century. The reasons for this are as follows.

In the first few years after the discovery Dubois was intensively occupied defending his ideas. He exhibited the fossils frequently, responded to criticisms, and provided supplementary information. There were direct confrontations between supporters and opponents during these years, and it was possible to form an opinion from a personal inspection of the fossils. It was in this period that both sides took up their positions.

In about 1900 Dubois backed out of the discussion and withdrew the opportunity for studying the fossils. Except on a few occasions he maintained silence over his find and no longer responded to others' publications, his last significant contribution to the discussion being produced in 1898. In 1900 he wrote an explanation to accompany his own 'in-the-flesh' reconstruction of the ape-man, but thereafter for the next twenty years he released almost nothing. Partly because of Dubois' attitude the debate reached calmer waters after the turn of the century; it lost its fierceness and those involved returned to their own affairs. When no new information was forthcoming the discussion became increasingly derivative. New interpretations only emerged from new discoveries of fossil hominid remains or from changes in the way of viewing the process of humanisation. A natural dividing line can thus be drawn across the debate, roughly coinciding with the turn of the century.

The publications from the years leading up to and including 1899 give a representative picture of the variety of ideas over Dubois' find. They enable us to approach our intended analysis of the reception and significance of his work.[1] A broad outline will suffice for the period following this, also because Dubois took only a minimal part in later discussions. In the next chapter we shall discuss the new impetus given to the debate by Black's and Weidenreich's excavations in China and Von Koenigswald's on Java.

Criticism

Even before Dubois published his official treatise in 1894 there were several reactions to his announcement of the discovery of a missing link. Scepticism prevailed. In 1893 the *Tijdschrift van het Aardrijkskundig Genootschap* gave a literal transcription of Dubois' description of *Anthropopithecus erectus* in his third quarterly report for 1892.[2] J.A.C.A. Timmerman, the Society's secretary, added a note to Dubois' final conclusion that the East Indies had been the cradle of the human race: "This last conclusion really does seem to have been made rather hastily."[3] Timmerman felt that Dubois should first have demonstrated that similar finds could not be expected outside the East Indies. He made no further comments.

The information in the geographers' journal attracted the attention of the German zoologist Paul Matschie, who summarised Dubois' results in the *Naturwissenschaftliche Wochenschrift* and then launched a fierce attack.[4] In his view Dubois' arguments that the bones belonged together were inadequate, and if they did in fact come from the same organism then it was clearly a human being; the femur provided the proof of this. Matschie voiced the suspicion that Dubois had not compared the skull with that of an orang-utan, and he concluded by asserting that Dubois' conclusions should not be taken seriously.

Dubois' comprehensive description of the fossils became available in Europe at the end of 1894, and criticisms followed thick and fast. One of the most detailed came from the Parisian anthropologist Léonce Manouvrier, who was later to stand forward as Dubois' most fervent champion. Manouvrier's first commentary, which he gave on 3 January 1895 at a meeting of the *Société d'Anthropologie*, was marked by extreme caution.[5]

Manouvrier began by pointing out that, considering the distance between molar, skullcap, and thigh bone when they were unearthed from the sediment, it was definitely not proven that they belonged together. In his view the femur was indistinguishable from a human thigh bone, and the abnormalities Dubois had indicated fell within the range of variation for human femurs. This did not exclude the possibility that the thigh bone had belonged to a non-human being, but the fossil in itself gave no cause for this supposition. According to Manouvrier the molar seemed to have come from an ape, perhaps a large gibbon or – if the ape were a recent one – from a female orang-utan. Yet alternative hypotheses, including Dubois' own, should not be ignored.

Manouvrier estimated the skull capacity at 900 to 1,000 cc. If Pithecanthropus had been a human being with a normal human height, this capacity proved that they were dealing with the remains of an imbecile. However, Manouvrier believed that the eyebrow ridges alone made it doubtful whether one could talk of a human being; they were even more strongly developed than in the Neanderthals, while the skull itself was considerably smaller than the Neanderthal skulls. Furthermore the back of the head was clearly ape-like, with a strong resemblance to that of a young gibbon.

Where the skull was concerned Manouvrier had no objections to Dubois' interpretation. It was somewhat improbable that the ape-like tooth belonged with the cranium because certain features of the skull suggested a jaw formation that seemed too light for the molar. In Manouvrier's view there were no formal objections to the idea that the three pieces belonged together, but the idea remained a hypothesis only.

There was still every chance that a low-status human race was involved. Manouvrier thought it unlikely that the remains were deformed but, once again, theoretically that possibility could not be excluded. In that case,

Pithecanthropus must have suffered from microcephaly but this disease only very seldom occurred. The assumption that the skull had belonged to an abnormally small but human individual would nevertheless gain in probability if it became clear that the femur did not belong with the skull.

Manouvrier ended his argument by indicating that Dubois' find could obviously have great importance for the theory of evolution, since the theory predicted the existence of transitional forms between ape and Man:

> The supporters of the creationist doctrine can content themselves... with quoting the maxim '*natura non facit saltus*'. This maxim is no cause for displeasure for the supporters of the evolutionary doctrine. For the latter, it would be no small reason for satisfaction if one came to conclude that, even where the human species is concerned, nature has proceeded as if she had been unwilling to pass from one species to another without producing an intermediary category.[6]

The French zoologist August Pettit also produced a carefully formulated critique.[7] In his view the material excavated was too incomplete to permit a definite verdict. The circumstances surrounding the discovery also required further clarification. For the time being Pettit thought it best to attribute the bones to a human. The vaulting and the measurements of the skullcap, and the torus occipitalis plainly pointed in that direction. True, the skull capacity as calculated by Dubois was very low, but considering his rather imprecise method of measuring one could not exclude the possibility that the capacity was considerably higher. For Pettit the features in which the molar differed from a human tooth were not a decisive factor, and the same applied to the femur.

The caution of Manouvrier's and Pettit's comments contrasts sharply with the assertive tone of the first responses from Germany. The German critics were particularly merciless towards Dubois' central proposition that he had found a transitional form.

The eminent anatomist Wilhelm Krause was the first to discuss Dubois' treatise in a thorough fashion. In a lecture given on 19 January 1895 at the *Berliner Anthropologische Gesellschaft* he allowed it to be understood in somewhat facetious terms that, as an anatomist, he would find it easy to make short work of Dubois' ideas.[8]

The skullcap was undoubtedly an ape's, probably a large gibbon's. In contrast the femur must have been a human's, thus excluding the possibility that the two items belonged together. This anatomy-based conclusion, continued Krause: "would remain valid, even if at the moment of excavation skullcap and femur had been found in contact with each other, [because in] river banks the most disparate things lie next to and on top of each other..."[9]

To illustrate his thesis that the skull came from a large species of gibbon, Krause pointed out the similarities of form and the absence of any trace of crests. He had no doubt that the molar was an ape's and thus could belong together with the skull. He gave greater attention to the question of the femur: "No descriptive anatomist in the world... would look upon this femur as anything different from a human thigh bone."[10] The abnormalities Dubois had mentioned were insignificant and also occurred as variants among humans. Furthermore these abnormalities could not be called ape-like, since they were not found in the orang-utan. The unavoidable conclusion to be drawn from the morphology of the thigh bone was that the rest of the skeleton, as well, could only have been human. That a being with a fully human body would have borne an ape skull was a mechanical impossibility, said Krause. He nevertheless decided that Dubois had made an important discovery. Besides bringing a new gibbon species to light he had also rendered acceptable the idea that humans had existed in the late Pliocene.

In the discussion that followed Krause's lecture, the anthropologist Felix von Luschan treated Dubois' conclusions with greater sympathy.[11] The resemblance between the fossil skull and that of a gibbon was not so obvious to him as it was to Krause, and he consequently considered that it was not a priori impossible that the pieces belonged together. In contrast the anatomist Wilhelm Waldeyer fully concurred with Krause's views on the skull.[12] He thought it possible though unproven that the thigh bone had belonged to the same organism.

Rudolf Virchow produced the most comprehensive contribution to the discussion.[13] He acknowledged the importance of the find, but like Krause he was unimpressed by Dubois' arguments. The brief description of the circumstances in which the discovery was made, was by no means sufficient to prove that the pieces belonged together. It would be foolish (*thöricht*) he thought, turning Dubois' thesis on its head, not to make a critical enquiry into whether the pieces were actually related.

Along with Krause Virchow also thought it likely that the skullcap came from a large gibbon. The skull index and the form as a whole were clear indications of this, and the back of the head was also ape-like. Virchow regarded Dubois' method of estimating the skull capacity as so unreliable that it was impossible for the time being to reach any conclusions about the volume.

Virchow was even less willing to pronounce a verdict on the molar. On the matter of the thigh bone, to a large extent he shared Krause's opinion that there were only negligible differences between the fossil femur and that of a modern human. In Virchow's view Dubois' conclusions on the fossil being's way of moving in no way implied that Pithecanthropus must have been an intermediate form. The gibbons' upright gait distinguished them from the other apes, but this did not necessarily mean that there was any question of kinship with humans, especially since the gibbons always needed their long arms to keep their balance when walking on two legs. The fossil femur could have belonged to a bipedal gibbonoid form – the extreme elongation of this bone even provided a concrete argument to make this proposition – but this did not mean that it must have been a transitional form between ape and Man.

More detailed investigation would reveal what Dubois had discovered, but a missing link was out of the question. Virchow decided that Dubois' argumentation proved on closer examination to be based on wild speculations unable to withstand criticism.

The first responses from the British Isles were also dismissive, but the verdict was totally different. The British view of the femur was by and large the same as the German, but in their eyes the skull was human rather than ape-like and Pithecanthropus must thus have been a human.

The palaeontologist Richard Lydekker wrote a short critique of Dubois' treatise in the 24 January 1895 issue of *Nature*.[14] While not contesting the proposition that the fossils had belonged to one organism, he did raise objections to Dubois' interpretation. In his view the remains were human in origin. The absence of crests on the skull proved that it could not possibly have come from an ape. The skull was probably that of a microcephalous idiot.

The anatomist Daniel J. Cunningham discussed the finds more fully in a lecture given on 23 January 1895 to the Royal Dublin Society.[15] Cunningham first dealt with the three fossils. The narrowness of the forehead section, the marked development of the eyebrow ridges, and the limited vaulting in the skullcap could be seen as ape-like features, but this was not to say that an ape was involved here. Cunningham commented that it had been a grave oversight on Dubois' part that he had compared the fossil skullcap only with ape skulls and not with skulls from members of modern human races, and he also criticised Dubois' dismissal of Neanderthal Man. As Huxley had established, the Neanderthals represented an inferior human race.

In Cunningham's eyes the profile of the skull bore a striking resemblance to that of a microcephalic. He also detected a clear similarity to the Neanderthal and Spy crania. Nevertheless it seemed unlikely that the skull came from a microcephalic, and the resemblance was explained by the fact that microcephalous abnormalities could largely be regarded as atavisms. Cunningham saw the Neanderthals as occupying an intermediate position between Pithecanthropus and the human: "by a series of easy and nearly equal gradations we are led from the fossil form up through the Neanderthal and Spy forms to the modern cranial arch."[16] The Neanderthals also filled an intermediate position with regard to skull capacity, 1,200 cc, and from this Cunningham drew the (rather singular) conclusion that: "The fossil cranium described by Dubois is unquestionably to be regarded as human. It is the lowest human cranium which has yet been described."[17]

Cunningham believed the femur to be human, or at all events more human than the Neanderthals'. Considering the distance between skullcap and thigh bone in the sediment where they were uncovered, he thought it improbable that the bones came from one individual. The molar certainly possessed several ape-like features, but not to the extent that it could not have belonged to a human. Among the Neanderthals, Australian Aboriginals and Negroes one also found teeth that differed markedly from those of western and northern races.

Cunningham's final verdict was: "The so-called Pithecanthropus is in the direct human line, although it occupies a place on this considerably lower than any human form at present known."[18]

On 13 February 1895 Cunningham gave a lecture along the same lines to the Anatomical Society of Great Britain and Ireland.[19] In the discussion that followed the lecture the anatomist Arthur Keith observed that nothing could be stated with any certainty about the age of the fossils, or about whether or not they belonged together.[20] The bones were discovered in the bank of a river that swelled to a raging torrent in the wet season. Because of this the banks could undergo a considerable degree of disturbance each year.

Keith agreed with Cunningham's views on the femur and skullcap, but he was not completely convinced that the skullcap came from a normal individual. The molar was more likely to have belonged to an ape than to a human. As Pithecanthropus' jaw had probably not been very heavy, he thought it doubtful that the molar belonged with the skull.

Some time later Keith revised his opinion on a number of points.[21] He claimed that closer study had served to convince him that Cunningham's views on both skullcap and molar were most probably correct. However, he did not share Cunningham's objections to the femur:

> It seems to me... highly probable that the frame of man reached its perfection for pedal progression long before his brain attained its present complex structure. If one conceives this probable or even possible, there is no hindrance to awarding the femur to the Bengawan woman.[22]

In Keith's opinion the fossils, like those of the Neanderthals, belonged to members of a primitive human race:

> taking these as samples of their race we can say with some assurance that man has not changed much since the Tertiary period of the earth's history closed... [F]or the purpose of giving us a clue to the human line of descent, the fossil remains at present known assist us not a single jot.[23]

The anatomist William Turner also produced a comprehensive commentary on Dubois' treatise.[24] In a lecture to the Royal Society of Edinburgh on 4 February 1895, he began by observing that it had by then been generally accepted that the Neanderthals represented a primitive ("wild") yet unmistakably human race, comparable to modern Australian Aboriginals. Like Cunningham, Turner suggested that Dubois had made an error in failing to compare his finds with similar primitive races. He believed that the length and width, the pronounced eyebrow ridges, and the minimal vaulting of the Javanese cranium frequently occurred in Australian Aboriginal skulls as well. Furthermore, it was not uncommon to find a skull capacity of approximately 1,000 cc among Australian Aboriginal women.

According to Turner, the fact that there was little wear to be seen in the molar was an argument against the tooth and skullcap having belonged to the same owner; the cranium was unmistakably that of an older individual. The molar could also have come from an orang-utan.

Where the femur was concerned, Turner suggested that Dubois had underestimated the variation in human thigh bones. The features which had led Dubois to reject the classification of the fossils as human were only accidental variations. Turner doubted whether one could talk of one individual, in view of the considerable distance between the finds and the fact that the bones had been carried down to a river from somewhere else.

Taking everything into consideration Turner concluded that the fossils, with the exception of the tooth, could have come from a human race comparable to the Neanderthals and Australian Aboriginals. "The existence of... a transitional form" he decided "is still a matter of speculation, and has not been placed on the basis of ascertained fact."[25]

The French anthropologist Paul Topinard declared his general agreement with Turner's conclusions in a review of his article,[26] and at one point went even further than his English colleague. He perceived not just a resemblance, but an identity between Dubois' fossils and the Neanderthals. "The matter is decided in our view... [T]he Neanderthal type existed in Java in Quaternary times... "[27]

The reaction of the American palaeontologist Othniel C. Marsh was conspicuously different from all these sceptical voices.[28] On 21 January 1895 he commented on Dubois' work in general but extremely complimentary terms. In his view the unearthing of Pithecanthropus was one of the most important discoveries since that of Neanderthal Man. Dubois had examined and described the fossils with great care, and his conclusions were convincing:

> It is only justice to Dr. Dubois and his admirable memoir to say here, that he has proved to science the existence of a new prehistoric anthropoid form, not human indeed, but in size, brainpower and erect posture, much nearer man than any animal hitherto discovered, living or extinct.[29]

Yet the unfavourable criticism continued to dominate. Paul Matschie continued to hold the opinion he had expressed before Dubois' treatise appeared, that there was no proof of the fossils having belonged together, and that they were probably human.[30] He did not consider that even the great age of the fossils had been established. Lydekker's judgement reinforced his conviction that the fossils were nothing more than pathologically deformed human bones.

The Dutch anthropologist Herman F.C. ten Kate also reached a negative conclusion, published in the *Nederlandsch koloniaal centraalblad.*[31] In view of the considerable distance between the finds, he professed a greater willingness to be regarded as foolish (*thöricht*) than to accept the idea that the pieces belonged together. Dubois' calculation of the skull capacity was suspect – his estimate was probably too high. There was no doubt that the skullcap and molar came from an as yet unknown large species of ape. In Ten Kate's opinion it was for the time being unproven that the very human-like femur belonged with the other pieces.

The Swiss anthropologist Rudolf Martin produced a detailed critique in the March 1895 issue of the journal *Globus.*[32] In broad terms he concurred with the views expressed by his English colleagues. He acknowledged the possibility that the molar, skullcap, and thigh bone could have belonged to one organism, but he felt that the documentation of the circumstances surrounding the find was totally unsatisfactory.

In the curve of the supra-orbital ridges the fossil resembled a human individual, in Martin's opinion. Dubois' estimate of the skull capacity urgently required closer examination. Nonetheless a capacity of about 1,000 cc was no exception among human beings, and a torus occipitalis could also be found among modern humans.

Martin, like Cunningham, thought that Dubois was mistaken in failing to compare the Javanese fossils with those of Neanderthal Man. Without further elucidation he suggested that the skullcap resembled that of the Neanderthals "in all essential points."[33]

The molar showed too little sign of wear to belong with the skull. Finally, the femur was completely human. Thigh bones among humans frequently showed the abnormalities which had induced Dubois to adopt a separate classification. Martin concluded with the statement that there were insufficient grounds for Dubois' claim to have discovered a transitional form. "With that" he stated "all the author's other speculations, which he believes uncover the whole 'path of the origin of man', collapse spontaneously." He believed that true scholarship was not well served by the erection of "airy speculative constructions."[34]

Martin's interpretation of the skullcap, like Cunningham's and Turner's, was diametrically opposed to the views held by Krause, Waldeyer, and Virchow. The latter put up a good fight, responding in a lecture given to the

Berliner Anthropologische Gesellschaft on 15 June 1895.[35] Virchow continued to maintain that the skullcap was ape-like, being totally unconvinced of the supposed resemblances to the human. He mentioned the marked constriction of the cranium just behind the eyebrow ridges as a deviation from the human skull that, up to now, had been ignored by virtually every critic. Such constriction never occurred in the human being, he claimed.

In his lecture Virchow also enlarged upon the exostosis on the thigh bone. The most plausible explanation for this excrescence was that it had resulted from an infection, possibly originating in the spinal column. The cause was thus an internal disease and not a wound caused by an external agent. Virchow believed that the complaint had almost healed, which provided an argument for regarding the thigh bone as human because the healing process must have taken a long time, demanding a great deal of care and devoted nursing. Also, the patient must have been rendered totally immobile for a long time. All this suggested that the victim must have lived in a human community. Nevertheless Virchow did not rule out the possibility that the thigh bone was that of an ape.

Finally one other authority supported Dubois: this was Ernst Haeckel, who associated himself to a large extent with Dubois' conclusions in his *Systematische Phylogenie der Wirbeltiere* of 1895.[36] On the subject of primate fossils he observed:

> Some of these are certainly of great importance, especially the skullcap of the Pliocene *Pithecanthropus erectus* of Java (1894), which really seems to represent the 'missing link' so eagerly sought for, in the chain of transitional forms.[37]

Dubois' human family tree also met with Haeckel's approval. His only reservation concerned the femur. For the present, he thought it better not to rule out the possibility that it came from a fossil human race.

It is remarkable that at this point Haeckel turned his attention to the Neanderthals, after having completely ignored them until then: "The similar diluvial skulls of Neanderthal and Spy, with their low forehead and their jutting eyebrow ridges, probably belong to this chain [of transitional forms], too."[38]

However, in his *Systematische Phylogenie* he also warned against overestimating the importance of these and similar fossil finds:

> In our view, compared to the importance of those positive facts which prove the historical succession of the main groups of our chain of ancestors, there is much less phylogenetical significance in these petrifactions, which are supposed to supply the particular evidence of the transformation of one group into another.[39]

These transitions were already sufficiently documented by means of comparative anatomy and embryology, "whether or not later palaeontological discoveries are going to uncover any 'intermediate forms'."[40]

DUBOIS' REPLY

With the exception of Marsh's and Haeckel's positive reactions, the first commentaries on Dubois' treatise varied from the reserved to the extremely critical. The picture we have of the criticism dealt with up to this point is reasonably close to the sketch of the reception given to *Pithecanthropus erectus* we usually encounter in the existing literature: there were doubts about whether the fossils were linked with each other; the femur was believed to be human; the skullcap was attributed either to an ape or to a human.

Yet the debate continued and became more complex as it progressed. New viewpoints came to the fore, positions on the subject were modified or revised, and there was increasing diversity of opinion. Now Dubois joined in the discussion, his contributions being to a large extent responsible for the further development of the debate. The earliest commentators had only Dubois' treatise as basis for their opinions; his verbal and written explanations and supplements to his first publication played an important role in later critiques. Because of this we shall first examine Dubois' reply and the supplementary information he provided.

Dubois' chance to answer his critics came in the autumn of 1895, during the Third International Congress of Zoology held in Leiden from 16 to 21 September. Dubois had returned to the Netherlands with his collection of East Indian fossils shortly before this, in August. On the last day of the conference, 21 September, he was given the opportunity to exhibit the Pithecanthropus fossils and to defend his interpretation of the pieces. A great deal of interest was shown in them; about a hundred people attended the session including such prominent figures as W.H. Flower, A. Milne Edwards, Marsh, and Virchow, who presided over the meeting.

Dubois opened the session with a lecture.[41] On the basis of the fossils themselves he explained once again how he had reached his conclusions. He acknowledged that his publication had been inadequate in several respects, and devoted special attention to the points which had received too little elucidation.

He now dealt far more comprehensively with the circumstances surrounding the discovery than he had done in his treatise. He described the geological formation in which the fossil fauna had been unearthed, and the Pithecanthropus find-spot. In a profile drawing of the site he reproduced the positions of molar, skullcap, and thigh bone in the sediment (Fig. 10). He stated that the layer in which the finds were made was completely intact, and that the

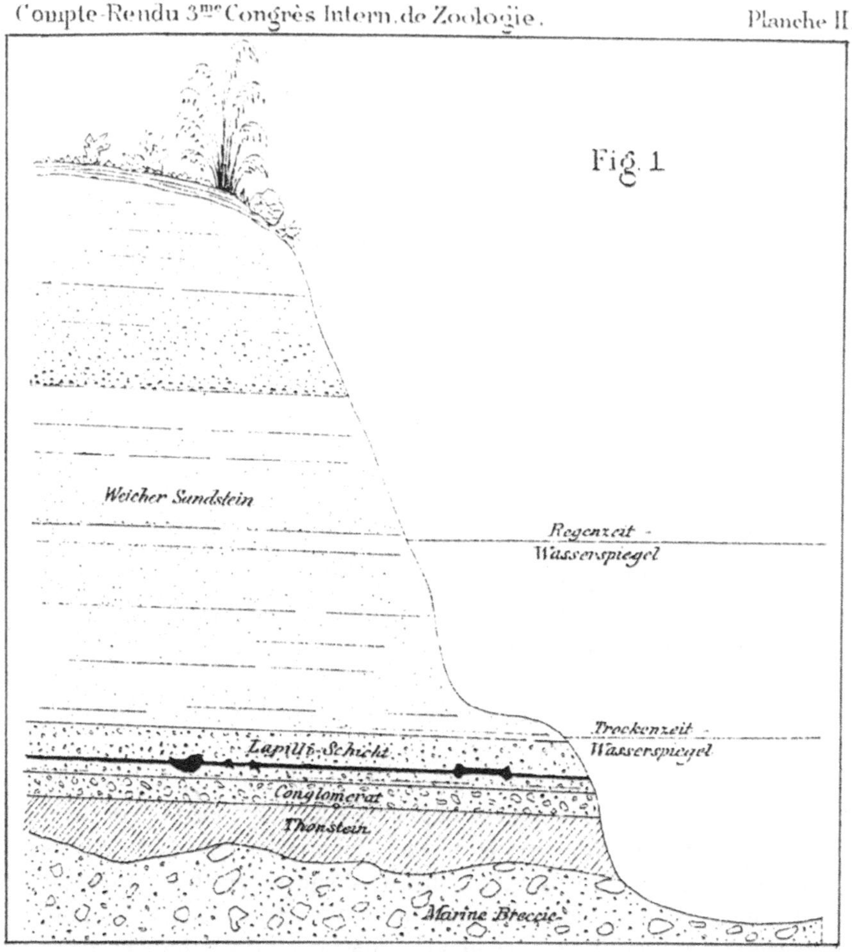

Fig. 10. Profile drawing of the bank of the River Solo at the Pithecanthropus site.

fossils showed no signs of having been transported from elsewhere. There was thus no question of their having been washed out of other layers. Further, the identical high degree of fossilisation among all the bones guaranteed that they dated from exactly the same era.[42]

Thus in Dubois' view the ground was cut from under any doubt, based on the circumstances of the find, that the pieces belonged together. It was not uncommon to find the bones of the same animal at a distance of anything up to forty metres from each other, in similar volcanic sediments. Dubois also provided data on the fauna found with the fossils, and explained how he had used the composition of this in arriving at a late-Pliocene or early-Pleistocene dating. However, for more detailed information on the fauna he referred his audience to his forthcoming publications on the excavations.[43]

Dubois admitted that his description of the ape-man fossils left something to be desired, but he stood his ground over his interpretation. Where the femur was concerned it could well be, that some of its abnormalities also occurred occasionally in humans, but he had been unable to find a human femur showing them in the same degree. The flattened form of the fossil's planum popliteum absolutely never occurred among humans. Among the 400 or more human femora he had examined with Manouvrier's help since returning from the East Indies, none was exactly like that of Pithecanthropus.[44]

Dubois agreed with his critics that all the principal features of the thigh bone had to be classified as human. For him this was a serious argument against Virchow's suggestion that the thigh bone could have come from a gibbon. The human characteristics of the femur were explicable if one accepted the idea that the bipedal gait had developed right at the beginning of the humanisation process. Thus, in spite of these human features it could well be that the thigh bone had not belonged to a human. Furthermore, the first 'real' humans appeared only in the middle Pleistocene.[45]

Dubois believed that his own interpretation was vindicated by the divergence in opinions on the skullcap. Where one expert could maintain that the skull had unmistakably ape-like features and another could point to obviously human ones, his own interpretation gained in plausibility with its incorporation of both views. He was adamant that classification either among humans or the great apes created inconsistencies, and that only the intermediate position did justice to all the skull's features.[46]

Replying to the criticism of his measurement of the skull capacity, Dubois pointed out that Manouvrier, the only one to have risked calculating the skull volume, had arrived at a similar estimate. He conceded Turner's and Martin's point that a skull volume of approximately 1,000 cc did occur in some human races, but in human beings this was always linked with a very small physical build. The fossil femur indicated normal human bodily measurements and the skull capacity was thus, relatively speaking, unprecedentedly low. Conversely in both relative and absolute terms the skull capacity was much too high for an ape.[47]

Dubois continued to maintain that the skullcap was ape-like in its general form; for this reason alone it could not be ascribed to a human. As an additional argument he pointed to the ape-like swelling of the parietal part of the skull, which Manouvrier had also spotted, indicating powerful development of the neck muscles.[48]

Dubois now also paid attention to the crania from Neanderthal and Spy – he had obviously abandoned his previous view that they were useless because of their pathological nature. He admitted that where form was concerned, there were clear resemblances to the Trinil skull, but he also perceived essential differences. The Neanderthal skull had both a higher vault and larger measurements and capacity than the Pithecanthropus skull.[49]

When discussing the molar Dubois introduced a new find. This was also a tooth, the second molar from the left side of the upper jaw. It was unearthed in October 1892, no more than three metres away from the skullcap. Dubois claimed that the piece had at first escaped his notice because it lay beneath a large number of other teeth dug up from the same place. This statement notwithstanding, it was probably the molar he had originally classified as belonging to a *Cynocephalus*, a baboon.[50]

In Dubois' eyes the circumstances surrounding the discovery and the morphology of the molar made it completely credible that this piece also came from Pithecanthropus. In contrast to the third molar this molar was badly worn and therefore more difficult to classify, but it clearly showed the same signs of retrogression.[51]

As in the case of the skullcap, commentators on the third molar differed in their opinions, from those who saw it as ape-like to those regarding it as human-like, and Dubois again considered this to be confirmation of his own interpretation. He disputed the view that the molar did not match the cranium because it showed so little wear: the third molars usually came into use much later than the other teeth, in apes as well as in humans.[52]

Dubois ended his lecture by concluding that the various criticisms of his treatise had given him no reason to revise his interpretation. He continued to maintain that *Pithecanthropus erectus* was a transitional form and that the differences between this form and both humans and apes were of a kind that justified the classification of the fossil in a separate family.[53]

No really new perspectives emerged from the discussion following Dubois' lecture. Marsh asserted that he had frequently come across disorders of the femur in apes comparable to the signs of disease in the fossil femur, and these had also healed. There were thus no grounds for Virchow's reasoning that the healing process could provide an argument in favour of the human character of the femur. The Utrecht anatomist Emil Rosenberg indicated that the form of the Pithecanthropus skull did not provide conclusive evidence of bipedal gait, since the basal part of the back of the head in New World apes was also bent forward to a marked degree. Finally Karl Martin observed that the Trinil layers could well be of late-Pliocene or early-Pleistocene age, because the marine sediment upon which the Kendeng formation had been deposited, clearly dated from Pliocene times.[54]

The debate ended indecisively.[55] The *Naturwissenschaftliche Wochenschrift* said that

> several misconceived opinions to which the first publication on the subject gave rise, have certainly been removed by these discussions, and in particular the nature and the age of the layers has been ascertained as far as

possible. A more or less certain and definitive answer to the question still lies however, in the future.[56]

Dubois did not let matters rest at this first rejoinder. He asked the chemist J.M. van Bemmelen to carry out age tests on several of the mammal bones found in the same layer as the Pithecanthropus fossils. For this purpose Van Bemmelen employed Carnot's fluorine method, based on the fact that the fluorine content of skeletal remains increases in proportion to their age. The results, published by Van Bemmelen in several articles, tallied reasonably well with Dubois' assumptions about a late-Pliocene or early-Pleistocene age.[57]

In 1895 and 1896, after his Leiden lecture, Dubois made the rounds of scientific institutions in western Europe, repeatedly exhibiting the fossils, and defending his interpretation of them with undiminished zeal. He visited Liège, Brussels, Paris (twice), London, Edinburgh, Dublin, Berlin, and Jena. Between trips he continued with his research and on several occasions added new elements to his argumentation.

In Brussels, where he addressed a meeting of the *Société Belge de Géologie* on 29 October 1895, Dubois discussed the circumstances of the discovery with several of those present, and for the first time displayed a map of the excavations in the banks of the Solo River.[58]

In Liège he studied the Spy fossils, now relinquishing the idea that the remains were pathological. He openly admitted his mistake in a lecture given in Dublin on 20 November 1895: "I am now wholly convinced that they are not at all pathological, and was much struck by the great resemblance with the cranium of Pithecanthropus."[59] Yet he immediately indicated once again what were, in his view, essential differences. The Pithecanthropus skullcap was more flattened in the parietal area, and had a relatively more strongly developed orbital area. Furthermore, the dimensions and capacity of the Neanderthal skulls were plainly larger.[60]

In Berlin Dubois devoted a good deal of energy to refuting the proposition put forward by Krause, Virchow, and Waldeyer that the skull was gibbonoid.[61] He argued on this point that any gibbon with the same measurements as Pithecanthropus would have had a much smaller skull volume. Using a series of examples he explained that brain size in large animals is always relatively less than in smaller species; in other words, body weight always increases more rapidly than brain weight. Then he showed that Pithecanthropus' body weight would have had to attain an absurdly high level to maintain the brain:body ratio typical of gibbons. Moreover, such a weight was not compatible with the features of femur and skull, which suggested a much lighter build. Dubois agreed with his German critics that the morphology of the skull was more gibbon-like than human-like, but Pithecanthropus was not a gibbon.[62]

In his Berlin lecture he was able to provide further details on the endocranial cavity, the internal part of the skull, which had been partially cleaned in the meantime. Among other things, he deduced from the sulcus transversus, now visible, which showed the border between the great and the little brain, that the foramen magnum in the ape-man was situated much further forward than in the apes, and probably even occupied the same position as in Man. It was thus clear from the skull, as well, that Pithecanthropus had walked upright.[63]

Dubois challenged Virchow's suggestion that the femur could have belonged to a gibbon. The thigh bone was not dead straight, lengthwise, as it is in the gibbon. Certainly, there was less curvature than the average for human femora, but it was not difficult to find human thigh bones which were equally straight. Neither did Dubois agree with Virchow's explanation of the exostosis; the form and position of the bone excrescence was not consistent with the cause Virchow had suggested. He exhibited a number of ape femora with comparable bone tumours that had also healed up.[64]

In Dublin and Berlin Dubois produced a graphic illustration (Fig. 11) of the family tree of *Pithecanthropus erectus*:

> I am well aware of the exceptional mortality of such trees, but I also know that parts of them at least often survive, from which new life emerges. One

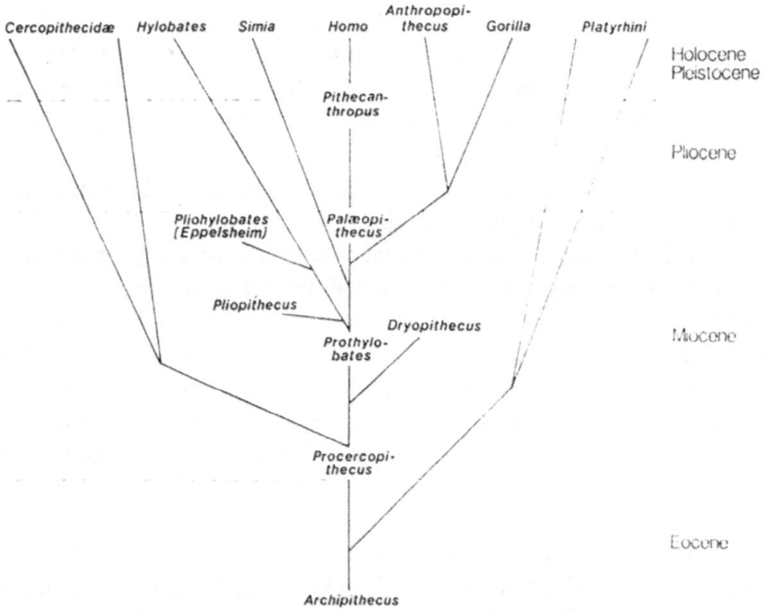

Fig. 11. Dubois' phylogenetic tree of the apes and Man (*Palaeopithecus = Anthropopithecus sivalensis*).

has to try to visualise the kinship relations of the forms presently known, and I know of no better means to this end than the form of a phylogenetic tree.[65]

Dubois was the first to realise and acknowledge just how 'mortal' his family tree was. As early as October 1896 he saw the need for a radical revision: *Anthropopithecus sivalensis* had to be deposed from its place as direct ancestor of Pithecanthropus. In 1895, while in British India, Dubois had studied the upper jaw of Lydekker's Siwalik chimpanzee and now, after making a careful comparison with living and fossil apes, he came to the conclusion that there could be no question of a strong resemblance to either chimpanzee or human. In his view an error in mounting the two parts of the fossil meant that a greater likeness to human proportions had been suggested than the piece in fact possessed. Some kind of anthropoid was probably involved here, but there was no notable resemblance to one of the living apes, and even less to Man; there seemed to be a greater resemblance to the monkeys. The name *Anthropopithecus* definitely had to go; Dubois suggested restoring the original name of *Palaeopithecus sivalensis*.[66]

It is clear that Dubois did not adopt a dogmatic stance on the exact genealogical position of the ape-man. We can see this from, for instance, a lecture he gave in Paris in June 1896 during the Quatorzième Conférence Annuelle Transformiste of the *Société d'Anthropologie*, which was devoted entirely to the Pithecanthropus question. In his lecture he stated: "After all criticism I persist in thinking that *Pithecanthropus erectus* belongs in a direct line to the genealogy of man or at least cannot be far away from this line."[67] In 1898 he expressed the same idea as follows:

> I cannot look upon *Pithecanthropus* but as upon a real member, the first known, in our genealogy. And even should we prefer not to regard him as a grandfather but as an uncle, still in every case he is a most venerable ape-man, representing a stage in our phylogeny.[68]

Here we can see clearly that Dubois' first concern was to establish the phylogenetic significance of Pithecanthropus as a transitional form, and that he attached only secondary importance to determining the being's exact position in the line of human descent.

In 1896 Dubois produced a more precise calculation of the skull volume in an article in the *Anatomischer Anzeiger*.[69] The sediment in the endocranial cavity had now almost been eliminated, making it possible to arrive at a proper calculation of the cubic content. Starting with the ratio of skullcap volume and total skull capacity valid for anthropoids, especially *Hylobates*, Dubois now calculated a volume of 900 cc or a little more.[70]

In the same article he provided the first reconstruction of the skull (Fig. 12). Its ape character is clearly visible in this, particularly in the teeth and

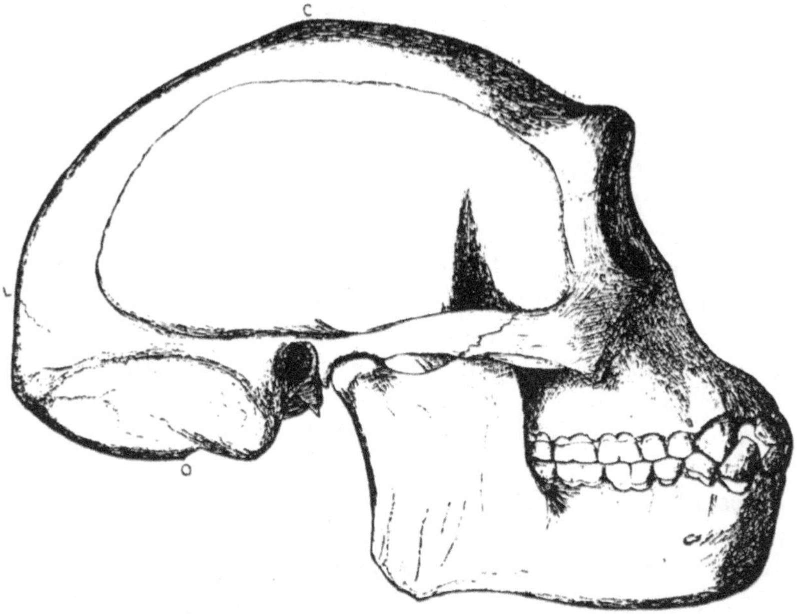

Fig. 12. Dubois' reconstruction of the Pithecanthropus skull.

the prognathism of the facial part. Once again Dubois emphasised the gibbon-like form of the cranium, and he deduced from this that the face must also have been gibbon-like. Apart from this, the reconstruction shows human-like features such as the more marked vaulting and the sharp bend in the lower part of the occiput. The whole thus displays a mingling of ape and human features, and illustrates in a striking way Dubois' belief that Pithecanthropus was neither ape nor human, standing somewhere between the two.[71]

During the Fourth International Congress of Zoology in Cambridge in 1898 Dubois once again announced a new find, this time a premolar from the lower jaw.[72] Kriele, continuing the excavations on Java after Dubois' departure, had found this in 1897. The tooth came from the same layer as the other Pithecanthropus fossils, "lying close to the place where the skull-cap and the two molars had been found, and exactly at the same level of the strata," Dubois said.[73] He believed that this fossil, in view of its intermediate character, also belonged indubitably to *Pithecanthropus erectus.*

In Cambridge Dubois for the first time showed a cast of the brain or rather, a cast of the endocranial cavity of the skullcap, and he gave a description of the brain on the basis of this cast. He deduced from the development of the lowest frontal brain convolution that there was already some capacity for speech. On the whole Dubois believed that the brain's characteristics, as far as these could be inferred from the cast, confirmed his thesis that Pithecanthropus occupied an intermediate position between Man and the anthropoids.[74]

Finally, we ought to pause for a while in front of the life-size reconstruction of the ape-man that Dubois made for the World Exhibition in Paris, in 1900 (Fig. 13).[75] The figure played no great part in the scientific discussion but it does provide a striking illustration of the way in which Dubois viewed his missing link. The fundamental idea underlying the reconstruction was that Pithecanthropus stood half-way between ape and Man.

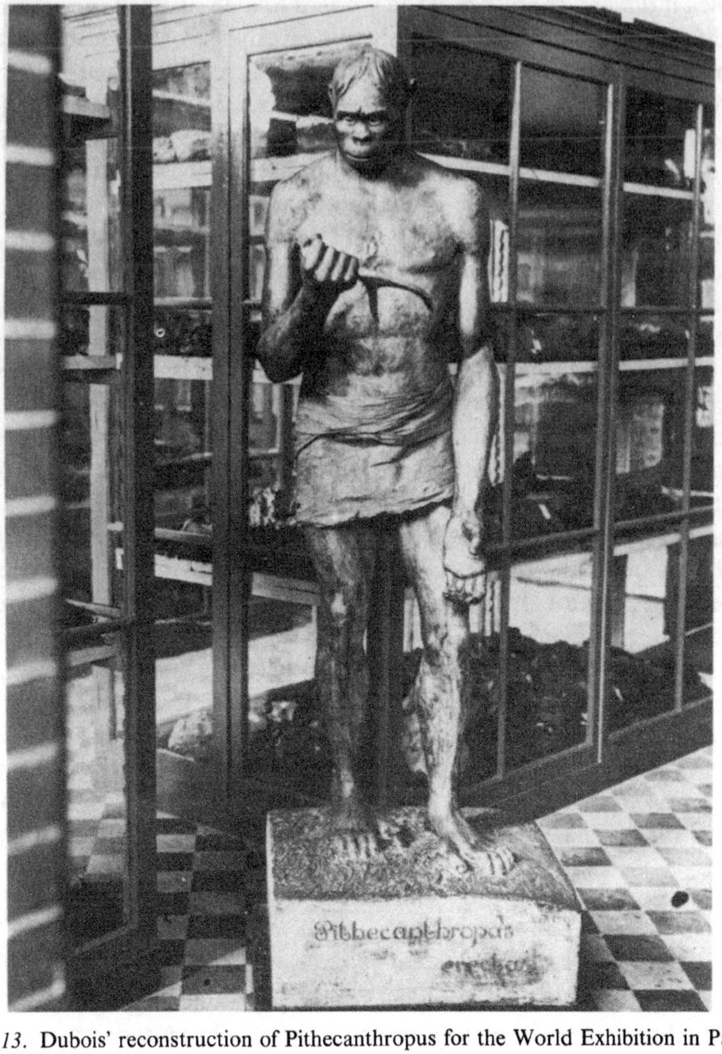

Fig. 13. Dubois' reconstruction of Pithecanthropus for the World Exhibition in Paris.

Dubois believed that the skullcap and particularly the teeth revealed an ape-like anatomy but also suggested development in the direction of Man. One can see this from Pithecanthropus' head. The facial part is prognathous but less so than in the great apes; the forehead recedes, as in the apes, but the high vaulting of the skull once again points in the direction of the human. Dubois also utilised the jaw fragment from Kedoeng Broeboes in his reconstruction, since in his view the piece revealed the somewhat more human development of the facial portion.[76]

The anatomy of the femur was human in its essentials, in contrast to that of the cranium, yet nonetheless several ape-like characteristics could be inferred from the thigh bone, according to Dubois. In his view the features in which it differed from that of the modern human justified the assumption that the legs, though suited primarily to a bipedal gait, were also adapted to climbing trees. He deduced from this the existence of a prehensile foot with an opposable big toe, somewhat separated from the other toes. The fossil's physical stance was the same as in the human, but the arms and hands must have been longer for climbing skills, with a relatively short thumb.[77]

By placing an antler in the ape-man's hand Dubois suggested that *Pithecanthropus erectus* had used tools. He also considered it likely that the ape-man had already begun to match the tools to the task at hand.[78]

The reconstruction also enables us to see how Dubois tried to stand up to the criticism of his views on *Pithecanthropus erectus*. To begin with he used the model to show that the combining of skullcap and thigh bone in one being was certainly not impossible, anatomically or mechanically; on the contrary, the result was an harmonious intermediate whole. Further he used the model to combat both the idea that Pithecanthropus should be regarded as an ape because of its skull morphology, and the suggestion that the thigh bone indicated a human being. Finally, the reconstruction was to make it clear that not only the combination of ape-like cranium and human-like thigh bone, but also each separate fossil piece, demonstrated the intermediate nature of Pithecanthropus.

On the other hand the reconstruction also reveals that the criticism of his views did not leave him altogether unperturbed. We can see this most clearly in his change of opinion on the limbs, which he had described as fully human in 1894. Yet he maintained – with undiminished force – his central thesis that Pithecanthropus occupied an intermediate position. The modifications he made in his original view were aimed solely at underlining this point.

Rejoinder

Several of Dubois' first critics once more pronounced judgement on the Pithecanthropus fossils on the grounds of the supplementary information he had provided. On the occasion of his tour of western Europe, especially, there were a great many reactions from researchers who up to then had remained aloof from the discussion. As we shall see, there remained a great division of opinion. Dubois succeeded in winning a number of experts over to his viewpoint, but new opponents were also surfacing. At the same time however, the discussion had clearly undergone a shift; although there was no consensus over interpretation of the fossils, the contrasts had become less extreme and Dubois' allies were increasing in number. Even more significantly, the fossils of other hominids, especially the Neanderthals, were being brought into the debate with increasing frequency. Because of the discussion surrounding Dubois' discovery, the problem of interpreting these earlier finds became topical once again. The consequences of this renewed interest were far-reaching, both for interpretation of Neanderthal Man and Pithecanthropus, and for palaeoanthropology as a whole. Let us now pursue the course of the debate from the point where Dubois joined in the discussion.

Manouvrier became one of Dubois' most dedicated champions from the moment when Dubois showed him the fossils at the end of October 1895, during a working trip to Paris. He suggested, in a "Deuxième étude" of the ape-man, the first part of which he published only a week after Dubois' visit, that: "The impression made on me by examination of the Pithecanthropus bones is very different from that made on me by the sight of Mr. Dubois' drawings, photographs, and descriptions, however exact."[79] The degree to which the bones were fossilised, and the supplementary information on the find-spot and the accompanying fauna, convinced him that Dubois was right in suggesting that the bones belonged together and that they dated from the late Tertiary or early Pleistocene. In his view the only remaining room for differences of opinion lay in the interpretation of the bones.

Manouvrier discussed the femur first. He had been unable to find a human femur exactly like that of Pithecanthropus. In particular, the form of the planum popliteum seemed to be extremely uncommon; he had only come across one comparable human femur, and that had been deformed. Yet Manouvrier insisted that the differences were too small to allow one to distinguish it from human femora with any certainty. An abnormality or a purely individual variation could not be ruled out but, Manouvrier hastened to add, this possibility in no way ran counter to Dubois' interpretation. In Manouvrier's view there could be no objection to attributing both molars to the owner of the thigh bone and skullcap. They resembled both human and ape teeth, and could thus be seen, rightly, as intermediate.

Manouvrier made a thorough comparison between the fossil cranium and the skulls of apes and humans, taking the Neanderthals into consideration as well. In his eyes the skullcap had an unmistakably intermediate character, and he illustrated this idea with a reconstruction of the skull (Fig. 14). This reconstruction creates a less ape-like impression than Dubois' model; even

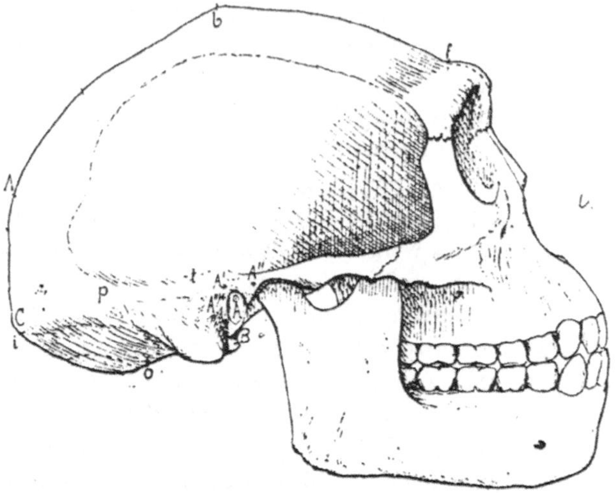

Fig. 14. Manouvrier's reconstruction of the Pithecanthropus skull.

though a mingling of ape and human features is discernible here, too, the prognathism is less pronounced and the teeth are more human. Nevertheless, Manouvrier observed in a later article that, in his eyes, Dubois' reconstruction was also acceptable and valid. He thought a more marked prognathism was in fact possible, because of the size of the teeth. On the other hand the skull's features, particularly in the temporal part, seemed to indicate a lighter jaw. Manouvrier emphasised that it was merely a question of the relative importance to be ascribed to the fossil's features; the phylogenetic position of Pithecanthropus was not at issue here.

Manouvrier devoted the last part of his comprehensive monograph to the question of descent:

> The scientific event, due to the laborious and fruitful researches of Mr. Eug. Dubois is of a nature to rejoice all the friends of science, but it seems that it must be more particularly agreeable to the transformists.[80]

The evolutionists, with whom he considered himself to belong, could quite justifiably regard Pithecanthropus as a powerful underpinning for their theory; the fossils had fulfilled a prediction made on the basis of the theory.

Manouvrier considered that Dubois' proposition that the ape-man was descended from a gibbonoid form was justified. An upright gait could have eventually been achieved if a gibbon-like form, already able to walk upright to a certain extent, had suddenly been forced to do so by a radical change in its living environment – deforestation resulting from a volcanic explosion, for example. The animal's anatomy would then have become increasingly better adapted to the upright gait as a result of increasing practice,

Manouvrier argued, clearly exhibiting a Lamarckian view of the evolutionary process. He even suggested an experiment with a group of gibbons: they should be put in treeless surroundings and watched closely to see if they walked upright more often, and if they became more efficient at it.[81]

The Pithecanthropus could have evolved subsequently from a gibbon fully adapted to walking upright and from this form the human, beginning with the Neanderthals, could have emerged:

> The examination... from the viewpoint of the theory of transformism... permits us to consider the race in question as a very probable forerunner of the human species and, with the same right as the race from Neanderthal and from Spy, as an ancestral form.[82]

The only point on which Manouvrier disagreed with Dubois was the latter's classification of *Pithecanthropus erectus* in a separate family. The ape-man had clearly travelled a great distance from the apes and represented a stage on the road to becoming human. In Manouvrier's view this justified a classification with human-like beings, and he saw little reason for defining a separate family.

In France De Mortillet also agreed with Dubois' conclusions.[83] As we saw in chapter 1, De Mortillet was the first scientist in France to propagate the idea of a transformist interpretation of fossil hominids. As early as 1879 he had outlined a hypothetical Tertiary transitional form between ape and Man, christening this being *l'Anthropopithèque*. He believed that his prediction had materialised in Pithecanthropus: "So Pithecanthropus is the intermediate [being] which links the gibbons or hylobatids with the lowest man known at present, the Neanderthal race."[84]

The zoologist August Pettit also subscribed to Dubois' ideas after the latter's visit to Paris,[85] and the anthropologist René Verneau supported him on the question of the skullcap and molars, although he believed that the thigh bone could only have come from a wholly human being.[86]

After having seen the fossils in Leiden, O.C. Marsh repeated his endorsement of Dubois' interpretation.[87] "[N]o palaeontologist who has had experience in collecting vertebrate fossils would hesitate to place them together" he argued,[88] and "Dr. Dubois... has met all the principal objections made to his views since he announced his discovery."[89]

Dubois also gained support from the Germans Nehring, Dames, Jaekel, and – as before – from Haeckel. The only objection raised by Alfred Nehring, zoologist and palaeontologist, was that it seemed unnecessary to classify Pithecanthropus in a new family.[90] The geologist Wilhelm Dames associated himself completely with Dubois' interpretation of the fossils, as did the palaeontologist Otto Jaekel.[91] Dames also endorsed the thesis that the ape-man must have emerged from a gibbonoid form, partly because the gibbons were the only extant species of primate with the ability to walk upright to a certain extent. Finally, Ernst Haeckel now gave his unconditional support to Dubois,[92] yet he clung to his belief that palaeontological evidence was only of secondary importance. He regarded the proof obtained through comparative anatomy and embryology as "much more important than if it had been possible to bring the whole series of our Tertiary primate ancestors, completely interconnected, before our eyes in a hundred fossil skeletons."[93]

However, Dubois also encountered opposition in Germany. Virchow held to his gibbon thesis,[94] believing that each fossil ought to be judged individually, and that the question of whether or not the fossils belonged together must not influence the verdict: "I have stuck to this point of view during all the debates in which I have taken part in the course of the years" he said. He had thus come to be convinced that the skullcap could only have come from an ape: "according to all the rules of classification this creature [was] an animal, to wit, an ape."[95] He now put forward the argument that the Pithecanthropus skull could be created by the linear enlargement of a gibbon's skull. Virchow also held to his opinion of the femur: it was impossible to decide anything definite about it, but one ought not to exclude the possibility that it had belonged to a large gibbon. He argued that the exostosis on the femur had been caused by an infection originating in the spinal column.[96]

Virchow felt that although Dubois' find was undoubtedly an important one, it provided no proof of the existence of a transitional form between man-ape and Man. Furthermore, evidence of descent from the apes would have to satisfy many stringent demands:

> That Pithecanthropus was a transitional form, will remain doubtful until it has been shown how this transition came about. For me a gibbon is as good a transitional form as Pithecanthropus: the latter has brought us nothing new, apart from its size... When [Dubois] assumes that each of the diluvial finds of humans has brought the human type closer to that of the ape, this amounts to mistaking individual, preconceived teleological notions for scientific evidence... Let us beware of weakening the public's faith in the certainty of scientific research![97]

Volz, Bumüller, Hubrecht, Branco, Kollmann, Klaatsch and others also defended the thesis that Pithecanthropus was an ape. The geologist Wilhelm Volz and the anthropologist Johannes Bumüller followed in Virchow's footsteps, believing the skull to be that of a gibbon-like ape.[98] The femur was indeed very human-like, but could nevertheless still have belonged to a gibbon. At best only the existence of a large gibbonoid, walking upright, had been established, but not that of a transitional form. Volz thought that certain finds of human bones and tools proved that there were already humans in the Tertiary, and therefore Pithecanthropus was in any case too recent to be eligible for the position of human forbear.

The zoologist A.A.W. Hubrecht reasoned along the same lines, also pointing to Tertiary-period finds; he could discern little in Pithecanthropus beyond an "East Indian ape species, a close relative of the gibbon."[99] Hubrecht also believed that there remained room for doubt about whether the fossils belonged together. In contrast the palaeontologist Wilhelm Branco wrote that "no one familiar with palaeontological research" could entertain doubts that the pieces belonged together.[100] He believed Pithecanthropus to be an anthropoid belonging to an unknown genus.

In the opinion of the anthropologist Julius Kollmann, *Pithecanthropus erectus* could not be an intermediary form because the process of humanisation had followed an entirely different pattern.[101] He thought that the human being had originated in a line of small apes, possibly gibbonoid or even lower. His chief argument in support of this idea was that larger forms had always developed from smaller ones. The pygmies were probably the ancestral form of humankind since, according to Kollmann, they had sprung from a small ape form. It was possible that Man had, as Dubois believed, developed from a gibbonoid form, but at best Pithecanthropus could be considered a failed attempt at becoming human.

The anthropologist Hermann Klaatsch thought that the human had originated in a hypothetical form, living in the Tertiary period, which he called *Proanthropus*.[102] This form had been more human-like than the present-day great apes. In Klaatsch's view the latter had stemmed from *Proanthropus* by a process of degeneration. Pithecanthropus could also be seen as a degenerate branch which, like the gibbon, had retained a number of the human-like features of *Proanthropus*. It was also possible that Pithecanthropus had belonged to a lower-level *Proanthropus* group that had eventually become extinct; at all events, it had not resulted in the human being. Thus Klaatsch felt that there was no place for Pithecanthropus in the human line of descent.

Rudolf Martin had originally been one of Dubois' fiercest critics, but his verdict was considerably milder after Dubois had clarified his thesis.[103] Martin did not agree with Dubois' interpretation, but now he felt that Dubois' conclusion could not be refuted in any definite way on the grounds of the fossil material available. The remains still appeared human to him, but he could no more produce indubitable proof for this than Dubois could for his conception of the fossils. For the rest, in his commentary Martin was especially opposed to Virchow's suggestion that the skull could be that of an ape, and stated his complete agreement with the arguments Dubois had brought to bear against this idea.

Conversely the Belgian anatomist and anthropologist Émile Houzé was sternly critical.[104] He began by making it clear that he wanted to base his views exclusively on the anatomy of the fossils: "I refuse to let myself be influenced by considerations concerning the sediment or age."[105] Houzé believed the femur to be fully human, and the same applied to the molars, which probably derived from a low-level human race comparable to the Neanderthals.

The skullcap had led Dubois to conclusions which could be dismissed as "arbitrary and fantastic."[106] It was impossible to orient the skull accurately, since the orientation points necessary for this were lacking; all profile comparisons were therefore pointless. Dubois' estimate for the skull capacity was not worth very much, either. Judging by the form of the skullcap, Houzé could only find similarities to such human races as the Australian Aboriginals and the Neanderthals. Hence the cranium was undoubtedly that of a human. In conclusion Houzé suggested that:

> Whatever the terrain where it was found, a bone which shows all the characteristics of a human [bone] must be considered as such. When, after determining this, it is said that it does not in the least weaken the hypothesis that the bone could have belonged to an intermediate species, one is abandoning the domain of facts without any plausible reason.[107]

Houzé's rather venomous article elicited a sharp reaction from Manouvrier, who now revealed himself as 'Dubois' bulldog'.[108] He upbraided Houzé for his extremely positivistic outlook. It was in fact both permissible and good sense to allow theoretical considerations a role in the interpretation of the fossils. It was precisely because Houzé refused to do so that the significance of the find had completely escaped him. If he had interpreted the similarity between Pithecanthropus and Neanderthal Man in an evolutionary context, then he would have reached an entirely different conclusion, according to Manouvrier: "Once the characteristics have been determined, the theory must intervene in their interpretation"[109] and "I prefer to believe that if the cranium from Trinil presents a form analogous to the cranium from Spy, with the exaggerations foreseen by the theory, this is simply because the theory conforms to reality."[110]

Manouvrier had nothing against regarding Pithecanthropus as human as long as the being was recognised as a forbear of modern Man, phylogenetically speaking. He thought that the difference from Man was certainly big enough to justify classifying the fossil as a separate species.

Now let us turn to the British Isles. William Turner insisted on holding to the idea, expressed in his first commentary, that the remains were human.[111] Yet he was impressed by the degree of fossilisation in the bones, and no longer doubted that they were of the same age.

Turner drew attention to a number of similarities between the skullcap and that of Neanderthal Man, "which everyone now admits to be human."[112] The humanity of the Javanese fossil was, however, less obvious. "Now that one has seen [the skullcap] there is more difficulty in coming to a conclusion," he admitted.[113] He felt that if the bones really belonged together, this proved that Pithecanthropus was a human: "the establishment of the human character of the femur would require us to regard the calvaria as also human."[114] And for Turner it was beyond any reasonable doubt that the femur was human.

The anatomist David Hepburn, who undertook an analysis of the femur, came to the same conclusion as Turner, that the thigh bone could only have come from a truly human being.[115] In his view it thus followed that there was no connection between the skullcap and thigh bone, or that the calvaria must also be attributed to a human.

The zoologist William Flower emphasised the strong likeness of the cranium to a gibbon's skull, but gave no definite verdict on it.[116] The anatomists A. Thomson and G.D. Thane also reacted with reserve.[117] They felt obliged to defer judgement in view of what they saw as inconsistencies in the features of the skullcap and thigh bone. New discoveries would have to prove that the combination of an ape-like skull with a human thigh bone really was possible.

Arthur Keith now associated himself with Dubois' conclusions, in essentials,[118] entirely abandoning his original scepticism. He accepted the age of the fossils and the idea that they derived from one species, but in view of the degree of wear in the teeth, indicating a young adult, he was not convinced that the pieces had belonged to the same individual.

Keith agreed completely with the genealogical position Dubois had assigned to the ape-man, but he did not agree entirely with the name Dubois had given the fossil being. In Keith's eyes there were two disparities between ape and human skulls: the greater capacity and the absence of ridges and crests in the human skull. In the light of these criteria Pithecanthropus was humanlike.

D.J. Cunningham now also ranged himself on Dubois' side on several important points, thus revising his previous opinion that Pithecanthropus was fully human. The report in *Nature* on his observations after Dubois' lecture in Dublin stated that: "when he was brought face to face with the actual specimen, he failed to see any decided and leading human feature, except its capacity of 1000 cc."[119] He even considered the occiput to be fully ape-like, where Dubois perceived human characteristics. To a certain extent he now also agreed with Dubois' thesis that the cranium was similar to that of the gibbon in form.

Cunningham's view of the femur remained unaltered: it could only have come from a modern human; there were not even any features of fossil or primitive races. It was thus unlikely that the femur had any connection with the skullcap. Later, Cunningham published a genealogical tree, based on his interpretation of the skullcap, in which he located *Pithecanthropus erectus* at some distance below the Neanderthals on the human line of descent.[120] Like Keith he thought that Pithecanthropus ought to be counted as a hominid.

Others who came to the same kind of conclusion as Keith and Cunningham were the anthropologists J.G. Garson[121] and A.C. Haddon,[122] the geologist W. Sollas,[123] the palaeontologist E.T. Newton,[124] and the anthropologist and archaeologist R. Munro.[125] Newton endorsed Dubois' suggestion that the ape-man had emerged from a gibbonoid form, and Sollas saw *Anthropopithecus sivalensis* as a possible forbear. Sollas and Newton, like Cunningham, also placed Neanderthal Man between Pithecanthropus and the modern human. Munro declared that "many fossil remains of man [are] intermediary links which marked different stages in the history of mankind."[126]

The last contribution to the Pithecanthropus debate to be mentioned here came from the Strasbourg anatomist Gustav Schwalbe.[127] In May 1896 Schwalbe gave a lecture on Dubois' fossils and from then on he was intensively involved with the question. He obtained a cast of the skullcap and in the spring of 1897 he visited Dubois so that he could examine the fossils for himself.[128] The report on his lecture he set out to write grew into a 225-page treatise published in 1899, and this dealt only with the cranium, and then only in part. Schwalbe hoped to investigate the skull capacity, the femur, and other aspects of the skullcap in later writings, but in the event these publications never materialised. Only his study of the femur, still incomplete, was published by Eugen Fischer in 1921, five years after Schwalbe's death.[129]

Schwalbe divided his description of the skullcap into fifteen chapters, in each of which he analysed one of its features. His working method was extremely thorough and comprehensive. He continually compared the fossil with a range of ape and human skulls, taking both apes and monkeys into consideration and working with a broad spectrum of human races. To provide one example: Schwalbe disputed Virchow's belief that the postorbital constriction in the Pithecanthropus skull was a pithecoid property on the basis of a comparative analysis of no less than twenty-two ape skulls and 350 human skulls. Several chapters of his dissertation developed into independent anthropological or morphological studies. For example, he devised various new craniometric indices, and produced critiques of the methods proposed by Guiseppe Sergi for classifying human skulls, of Johannes Ranke's anti-evolutionary anthropology, and of Ernst Haeckel's and Max Schlosser's views on the phylogeny of the primates.

According to Schwalbe his examination of the Pithecanthropus fossils immediately made it clear to him that comparison with Neanderthal Man was vital. He had therefore studied this fossil hominid thoroughly, and included the results in his comparative analyses. In his introduction he explained that he had originally thought the Javanese cranium and the teeth to have come from an ape, and the femur from a human. However, Dubois' later writings and the verdict of palaeontologists and geologists such as Jaekel, Nehring, and Dames, had made him realise that there must be a link between the bones. His comparative study had also led him to a different interpretation of the fossils.

Schwalbe's results were the recording of eleven features in which the skullcap differed from that of *Hylobates* – "fully sufficient" he felt "to reject all attempts to designate the Pithecanthropus skull as that of a large Hylobates."[130] The differences from the orang-utan and gorilla were also obvious. There was a slightly greater resemblance to the chimpanzee, but on the other hand there were unmistakable disparities – the ape-man's characteristics were closer to the human. Comparison with the apes resulted in the realisation that Pithecanthropus, in spite of a certain similarity in form to both groups, could not be included among the apes of either the Old World or the New. Compared with the apes, the fossil skullcap:

> [occupies] a kind of mediating position, [it] avoids extremes, leads naturally back to forms from which a further development into each of the three groups of living simians would be understandable. Compared to the simians, Pithecanthropus thus turns out to be a more generalised form.[131]

From comparison with Neanderthal Man, Schwalbe concluded that Pithecanthropus came between that being and the apes. In some of its features the Javanese cranium was closer to the Neanderthals, while in others it bore a greater resemblance to the apes. Anticipating his intended publication on the skull capacity, Schwalbe established that the brain volume clearly separated Pithecanthropus from the apes. In his view the distance between ape-man and Neanderthal Man was shorter than that between the Neanderthals and Man. His final conclusion was that:

> If... the existence of a species of the genus Homo, different from all races of recent Man [i.e., Neanderthal Man], has been proved, this species and not Pithecanthropus should be considered as intermediate between Man and ape, while Pithecanthropus in its turn [occupies] an intermediate position between Homo Neanderthaliensis and the most highly developed apes.[132]

Even though Schwalbe remained noncommittal here about the Neanderthals' status, he was soon to come to a definite decision that they ought to be considered a separate human species. He extended the research on the

Neanderthals he had been carrying out during the course of his study of Pithecanthropus, to a comprehensive comparative analysis, the results being published in 1901.[133]

In this treatise Schwalbe dealt the final death blow to the thesis that the fossil remains of this hominid were pathologically deformed, and he allotted the Neanderthals a separate species, *Homo primigenius*. In his publications on Pithecanthropus and Neanderthal Man, in which he wanted to keep to the purely systematic-zoological, he did not touch upon questions of descent. Yet it is clear from other writings in which he did deal with the question of evolution, that he accepted a phylogenetic developmental line from *Pithecanthropus erectus* (or a closely related form) to *Homo sapiens* via *Homo primigenius*.[134]

FROM PITHECANTHROPUS TO AN EVOLUTIONARY PALAEOANTHROPOLOGY

On the basis of the data brought together in the three previous sections I shall now analyse the background to, and the development of the Pithecanthropus debate, and consider the significance of Dubois' find for the development of palaeoanthropology. For ease of survey, I shall first deal with several elements of lesser importance in the discussion, which can be disregarded hereafter.

Firstly, we can observe that the debate was concentrated on the skullcap and thigh bone. Views on the molars were also extremely varied, but the verdicts on the teeth were not of vital significance in the formation of opinion on Dubois' find as a whole. Important conclusions were all based on the femur and skullcap.

Secondly, the details of the find-spot were only a matter for discussion in the first phase of the debate. The first reactions to Dubois' 1894 publication on Pithecanthropus show that most commentators were far from satisfied with his brief statement on the circumstances of the discovery; only a few critics accepted the idea that the pieces belonged together before he had provided more details. After Dubois joined the debate the outlook became more favourable, with more than half the commentators then accepting, either explicitly or implicitly, the connection between the specimens. Most others regarded the matter as still undecided but they did not dispute the possibility of the connection. Those who did raise objections, Cunningham for instance, did so on the basis of the anatomy of the fossils rather than on the circumstances surrounding the find.

It is remarkable that researchers familiar with the palaeontological, archaeological, or geological fields were more inclined to accept the idea of the connection between the fossils than were those from other disciplines. The critics from these three fields in fact all accepted the idea; it was only the zoologists, anatomists, and anthropologists who objected to it. Several statements by authors from the various disciplines underline this tendency. For example, Marsh and Branco thought that no experienced palaeontologist could doubt the connection between the fossil pieces, and Nehring even declared that it was the palaeontologist's duty to establish this connection. In contrast Krause, Virchow, and Houzé wanted, in so many words, to avoid being influenced by the circumstances of the find, and to base their verdicts solely on analysis of the bones.

Discussion of the find-spot as such remained largely confined to the first wave of criticism, after which commentators concentrated on analysis of the fossils. In what follows we can therefore restrict our attention to ideas about the skullcap and thigh bone.

The influence of the anthropological theories and methods discussed in chapter 1 was unequivocally present in a number of the initial responses to Dubois' find. In the nineteenth-century anthropological context of interpretation, fossil hominids were visualised as races of modern Man; the fossils were granted only taxonomic significance, not a phylogenetic one. This outlook is clearly reflected in a number of researchers' opinions on the cranium. Martin, Cunningham, Turner, Keith, and others attributed it to a human because in their view the skull's features fell within the scale of variation for a *Homo sapiens* skull. They considered the skullcap to have come from a low-level race, but all of them emphasised that there was no question of an intermediate form between ape and Man. Lydekker also thought of a human being, but he believed the anomalous features could be explained as pathological deformities. In short, Pithecanthropus was treated the same way as Neanderthal Man years earlier. Turner, Martin, and Cunningham also cited the Neanderthals in their argumentation, and the last two also brought the Australian Aboriginals on stage. According to Topinard Pithecanthropus and Neanderthal Man were identical, and Pithecanthropus therefore belonged to a primitive human race. It is hardly surprising that all these critics regarded the thigh bone as completely human.

Cunningham appears to be the only one to have harboured doubts about whether or not Pithecanthropus could be considered an ancestor of the modern human. His comments on this point are ambiguous; it cannot be inferred with any certainty from his article whether he saw Pithecanthropus as a member of a low-level race, or whether he regarded the being as a potential forbear of the modern races. In Manouvrier's first commentary, in which he associated himself closely with the English reactions, the possibility of an ancestral form was in fact explicitly recognised.

The German critics who attributed the remains to an ape were, methodologically speaking, closer to Dubois than the English who had constructed a human being from the fossils. Virchow, Krause, and Waldeyer followed Dubois in his morphological approach, placing the fossils a long way outside the spectrum of human measurements and forms. Like Dubois they concluded, from a clearly typological interpretation of the cranium form, that ape-like characteristics predominated. Yet because they neither accepted Dubois' contention that calvaria and femur belonged together nor trusted his estimate of the skull capacity, they concluded that the skullcap in fact came from an ape.

It is certainly no coincidence that it was the Germans, especially, who regarded Pithecanthropus as an ape. Virchow had great authority, and we can see how heavily his judgement weighed from the fact that the Neanderthals were not taken into consideration by most German scientists; after all, Virchow had claimed that only the disease-distorted remains of completely human individuals were involved. This view also influenced interpretation of the Javanese fossils. For the Germans there was no 'bridge', in the form of the Neanderthals, between Pithecanthropus and Man, and consequently there was less of a tendency to place Pithecanthropus in the same class as humankind, as the British were doing. In fact Dubois himself had argued along the same lines in his 1894 treatise.

Naturally, the wide scope for interpretation offered by Dubois' first description of the fossil pieces contributed greatly to the extremes of opinion evident in the first critiques. Yet it was the methodological approach that ultimately determined the use made of this room for manœuvre and the conclusions drawn.

The picture changed once Dubois had provided supplementary information and had displayed the fossils to several audiences of experts. The sight of the actual specimens clearly made a stronger impression than Dubois' illustrations and descriptions had done. For instance, Cunningham and Turner said that the skullcap was in fact much more ape-like than they had been able to infer from Dubois' treatise. In contrast Schwalbe now repudiated his original view that the skullcap was that of an ape. The degree of fossilisation convinced virtually everyone of the fossils' great age, and consequently the number of Dubois' supporters grew. Yet the opposition in no sense faded away entirely; in Virchow's circle especially, there continued to be a great deal of resistance to Dubois' ideas. I shall first discuss Virchow's position in some detail, after which I shall examine the new direction taken by the debate.

Virchow's comments fit in neatly with what we know about his ideas on scientific method.[135] He certainly cannot be called an anti-evolutionist, but he did have serious objections to an evolutionism à la Haeckel and Vogt. In his eyes these researchers came close to lapsing into the speculative approach of the *Naturphilosophen*. In his youth Virchow had fiercely resisted *Naturphilosophie* and had championed a strictly empirical and inductive approach to scientific problems. More than once he called himself a 'friend' of evolutionary theory, but he felt that the hypothetical nature of the theory should be kept in mind. He believed that the empirical foundation of the theory was as yet so unstable that it was premature to confirm its validity; Virchow demanded proof.

In the years that followed, his attitude became less and less reasonable and finally extremely rigid. There is no doubt that a vehement controversy with Haeckel, bearing political overtones, contributed to this.[136] His demands for material evidence took ever more exorbitant forms, and he seemingly lost sight of the fact that irrefutable proof does not exist. He accepted none of the evidence laid before him, and he was obviously unaware of the sometimes clearly ad-hoc nature of his arguments.

The request for more material proofs was also the main focus of interest in Virchow's articles on *Pithecanthropus erectus*. He repeatedly emphasised that no one had as yet produced the proof that Pithecanthropus was a transitional form. For Virchow, only a complete chain of transitional forms between ape and human would suffice as proof. He also continually pointed out that the material gave no reason whatsoever for supposing that an intermediate form was involved. The skullcap was plainly ape-like, and even the thigh bone could be fitted into the view that Pithecanthropus was an ape.

It is impossible to determine whether Virchow himself believed in his gibbon thesis, or whether he was only concerned with disputing the transitional character of the fossils. If the latter were true it would seem to me – assuming for a moment that such tactical considerations played a part in Virchow's judgement – hardly any wonder that he should stamp 'ape' rather than 'Man' on Pithecanthropus. If he had emphasised the human character of the fossil he might have found himself in considerable difficulty over the interpretation of the Neanderthal. Various commentators had drawn attention to the resemblances between Pithecanthropus and the Neanderthals, and Virchow would have had to provide himself with several new ad-hoc arguments in order to give a plausible explanation for these similarities without having to revise his verdict on the Neanderthals.

We should not lose sight of the fact that there was also a positive side to Virchow's standpoint. In certain cases his demand for proof was perfectly justified. For instance, he was the only critic to register an objection to the assumption of kinship between humans and gibbons on the basis of the so-called upright gait of the latter; he correctly pointed out that this was pure speculation. None of the other commentators discussed here took advantage of this justifiable criticism.

Johannes Ranke's interpretation was entirely in accordance with Virchow's. Ranke was a personal friend of Virchow's and a kindred spirit; he, too, perceived the evolutionism of Haeckel and company as a retreat into *Naturphilosopie*. Apart from this he was a devoutly religious man.[137]

Although he did not openly anathematise the application of evolutionary theory to Man, he welcomed any counterargument, if only to be able to take the wind out of Haeckel's sails. It is thus hardly surprising that he thoroughly approved of Virchow's reasoning and his judgement. The same holds for Johannes Bumüller, a Catholic priest, who was asked by Ranke to corroborate Virchow's analysis of the femur.[138]

There is greater difficulty in demonstrating that Virchow had any direct influence on the other supporters of the ape thesis. Yet whether or not his authority affected their judgement, other factors also certainly played their part. Volz and Hubrecht believed that there were already human beings in the Tertiary, and Branco made the same kind of suggestion. For them Pithecanthropus was simply too recent to be considered as a forbear of Man. Kollmann and Klaatsch also held opinions about human origins that excluded the Javanese primate as Man's ancestor, believing that the fossil could only have belonged to a side branch.

Although these researchers have to be counted among Dubois' opponents they cannot be placed in the same slot as Virchow and his supporters. In their publications we can see obvious signs of a shift in the debate. Kollmann and Klaatsch, particularly, showed a readiness to adopt a transformist interpretation of the fossils. Although they did not acknowledge Pithecanthropus as a forbear of Man they did observe, in so many words, that the fossil could be regarded as a being which had developed in the same direction. For them *Pithecanthropus erectus* was, so to speak, a failed attempt at becoming human, and they agreed with Dubois that they were concerned here with an intermediate form between apes and Man. Like Dubois they attributed a phylogenetic significance to the fossils' morphological features.

The kind of turn the debate took can best be illustrated by the responses from the scientists who revised their judgement on Pithecanthropus after they had actually seen the fossils for themselves, and after Dubois had clarified several points. Then Manouvrier, Keith, and Cunningham recognised that they were dealing with a transitional form linking humans with their ape-like ancestors. With this they abandoned their original framework of interpretation and accepted Dubois' belief that a phylogenetic significance must be assigned to the fossils' characteristics. This shift is particularly noticeable in the articles by Manouvrier and Cunningham. At first, Manouvrier hesitated between a taxonomic interpretation (as a separate race) and a phylogenetic one, and Cunningham's opinion was ambivalent on this point, although the taxonomic view predominated. In both scientists' later articles, however, the phylogenetic approach was adopted.

Over half the authors we have discussed chose this approach, meaning that for the first time a group of researchers, including internationally recognised authorities, acknowledged a fossil hominid as a transitional form, thus interpreting it in phylogenetic terms. In this context, it is of relatively little importance that only a few scientists agreed with Dubois' classification of Pithecanthropus in a separate family. The crucial factor is that they showed themselves ready to adopt an evolutionist interpretation.

From the reactions of some opponents we can also see that the influence of the nineteenth-century anthropological context of interpretation was waning. When he saw the fossils, Turner admitted that he now found it considerably more difficult to see the skull as completely human; doubt had set in, but he made no definitive pronouncement. Neither was there a verdict forthcoming from Martin, who nevertheless took a considerably milder attitude towards Dubois' viewpoint in his second paper. Houzé was one of the few who, having seen the fossils for himself, nonetheless still remained convinced that the characteristics of Pithecanthropus fell within the scale of variation among human races.

One can gather from the responses of several of Dubois' opponents that the humanity of the thigh bone was their greatest stumbling block. It seems as though some authors – Turner, for instance – would have had less trouble in cutting themselves free from the taxonomic interpretation if Dubois had found only the skull. These doubts about the femur are in fact quite understandable. Cunningham and Verneau could not bring themselves to ascribe the thigh bone to Pithecanthropus, either, and right up to the present day, some anthropologists object to the idea that the skullcap and thigh bone are connected, precisely because of the human appearance of the latter.[139]

As in the case of Virchow and Klaatsch, the views on Pithecanthropus expressed by Dubois' other critics were of course also influenced by their attitude towards evolution theory and the process of humanisation, while methodological and theoretical considerations as well as personal factors also played a part. I shall not delve into this in any detail, but only provide a few examples as illustration. It was not at all strange that Marsh and Haeckel should have accepted Dubois' conclusions with such remarkable celerity, given their unconditionally evolutionist outlook. They had both accepted and defended the existence of missing links long before Dubois' discovery, and one could say much the same for Manouvrier and De Mortillet. The Paris *Société d'Anthropologie* provided a platform for evolutionist discussions within a world of French scientists who had always been sceptical.[140] Consequently Manouvrier leapt at the chance to use Dubois' finds as 'propaganda' for evolution theory in his articles. De Mortillet also held up Pithecanthropus as proof of theories he had been defending, against the stream, for years. Houzé did not oppose evolution theory, but he opted for an extremely inductive approach, comparable to Virchow's. Finally Turner was known for his extreme caution and aversion to hypothetical reasoning; "caution was his watchword" as one biographer put it.[141]

There is a second element in the Pithecanthropus discussion indicating a shift in interpretation, one that has been implicit in what we have written so far, but which now deserves to be brought to the fore. The Pithecanthropus debate also revived interest in other fossil hominids, particularly the Neanderthals, and various authors now looked at these from an evolutionist perspective. Because of the resemblances between Pithecanthropus and the Neanderthals the possibility of an analogous interpretation presented itself almost automatically. Neanderthal Man took up a mid-position between the ape-man and modern Man, and the idea of a phylogenetic sequence was virtually self-evident.

Authors who opted for such an interpretation usually did so with what looks like some hesitation, or at all events without further explanation. Manouvrier actually stated in so many words that he saw Neanderthal Man as a transitional form between Pithecanthropus and the modern human, but the opinions of Cunningham, Sollas, and Newton on this matter remained much more implicit. In contrast the genealogical trees drawn up by Cunningham and Sollas leave no room for doubt that they supported the phylogenetic interpretation.

If these authors, with their cautious approach, show the stir caused within the traditional interpretive framework by the Pithecanthropus debate, Gustav Schwalbe's work broke right through it. In his writings he defended the new viewpoint in all its implications, and adopted an unconditionally phylogenetic approach to both Neanderthals and Pithecanthropus.

Schwalbe published his study of Pithecanthropus as the opening contribution to the *Zeitschrift für Morphologie und Anthropologie* appearing for the first time in 1899. He had edited this journal under its previous name of *Morphologische Arbeiten*, and continued as editor of the new version. There can be no doubt that Schwalbe's involvement with *Pithecanthropus erectus* and other fossil hominids was the immediate impetus to the change of course taken by the journal. In an editorial introduction to the first issue, entitled "Ziele und Wege einer vergleichenden physischen Anthropologie" (Aims and Methods of a Comparative Physical Anthropology) Schwalbe pleaded for an anthropology based on morphology. The aim of the new journal was:

> to contribute to the important question of the origin of the human kind by careful comparative anatomical and developmental investigations, and to investigate the relationships between the human races, their bodily development and their higher or lower position via a purely morphological approach.[142]

Schwalbe allotted a central place to palaeontology, and he viewed both mammalian palaeontology and palaeoanthropology as indispensable for the study of human phylogeny:

> The apparently wide chasm which separates Man and the simians can only be bridged by a consideration of the fossil forms which have only recently become known. A thorough study of the fossil remains of the order of primates... is an absolute prerequisite for the foundations of zoological anthropology.[143]

Among the examples he gave of such studies Schwalbe mentioned Dubois' work and his own forthcoming treatises on Pithecanthropus and Neanderthal Man. At the end of his introduction he returned once again to the great importance of palaeontology:

> In this way palaeontology becomes the principal guide for our attractive field of research and thus, at the end of... our investigation of the aims and methods, we arrive at Cope's saying that "The ancestry of man is a question to be solved by palaeontology."[144]

Schwalbe's programme helped to place the phylogenetic approach to fossil hominids on a firm footing. He contributed no new theoretical insights; as he himself indicated, Haeckel and Huxley had already stressed the importance of a morphological approach to the problem of human descent. Yet no one had until then put the approach into practice. Within the prevailing anthropological framework of interpretation the question of a phylogenetic interpretation of fossil hominids did not arise, and palaeoanthropology was neglected by the morphologists. Schwalbe's work was a milestone because he suited the action to the word and actually established palaeoanthropology on a phylogenetic and morphological basis.

It is obvious how important Dubois' work was in this. Pithecanthropus gave the first encouragement to researchers to tackle a phylogenetic interpretation of fossil hominids, with Dubois himself setting the initial example. In a way the Neanderthals were rediscovered because of the debate on the ape-man, and Dubois contributed indirectly to this rediscovery. In the period under discussion he expressed no opinion on the Neanderthals' phylogenetic position; his only concern was the status of *Pithecanthropus erectus*. Yet it was his pioneering work that gave the impetus to the revision of the traditional notions. In my view this is the greatest contribution made by Dubois' work to the development of palaeoanthropology.

Erickson, in his *Origins of Physical Anthropology*, recognises the value of Dubois' studies of Pithecanthropus and Schwalbe's of the Neanderthals, in providing a vital impulse towards a phylogenetic interpretation of fossil hominids.[145] However, the link between the work of these two has escaped Erickson, who describes Schwalbe's research as being carried out independently from that of Dubois. Neither does he indicate how and why the old interpretive framework was gradually dismantled.

This can be put down to the fact that Erickson has not analysed the discussion of the ape-man, basing his description of this chiefly on secondary sources. Partially as a result of this, his view of the debate is too simplistic and some of it is based on incorrect data. Erickson ignores virtually all the nuances of the different researchers' opinions, classing Cunningham and Houzé together, for example, since both indicated similarities between Pithecanthropus and Man.[146] He ignores the fact that Cunningham leaned towards a phylogenetic interpretation while Houzé did not. Furthermore, he takes it for granted that researchers writing about Pithecanthropus either as a human or as an ape were all opponents of a transformist interpretation – quite wrongly, as I have shown.

In view of this it is not surprising that Erickson has concluded that Dubois had very little support, since he has completely overlooked those researchers who were in fact prepared to embrace an evolutionist reading of the fossils, without sharing Dubois' views on the phylogenetic position of Pithecanthropus.

Further, Erickson's statements about Manouvrier are misleading: "[he] expanded Dubois' arguments into a fully transformist scheme" and "[he explained] Pithecanthropus' peculiar morphological characteristics... in terms of evolutionary process."[147] In fact Manouvrier, like Dubois, mainly followed the morphological tradition in his interpretation of the ape-man, and in his arguments we do not find him, any more than Dubois, using such qualifications as 'ape-like' and 'human-like' in terms of developmental processes. Manouvrier only talked of evolutionary processes and their underlying mechanisms when he brought up the hypothesis that deforestation could have been the reason for the development of an upright gait. Yet Erickson does not produce this example; he apparently has something else in mind, probably the fact that Manouvrier always took the theory of evolution into consideration in his analyses. However, the explanation for this must be sought rather in the national scientific context.

In France the plea for evolution theory was decidedly necessary, especially where such a loaded subject as human descent was concerned. Even though the theory was gaining a more positive reception than over the preceding decades, discussion in France had certainly not ended. It is no surprise that a supporter of evolution theory would vigorously draw attention to his viewpoint and defend it whenever a good opportunity for doing so presented itself. The majority of researchers in Germany and Great Britain shared Manouvrier's view of evolution theory, but they considered it unnecessary to declare this in so many words. They switched directly to the more specific question of whether Pithecanthropus proved the descent of human from ape; they accepted the theory as a context that spoke for itself, requiring no further explanation and certainly no apology.

My chief objection to Erickson's account of the debate is that he takes no notice of the way in which it developed over time. He skates over the fact that the discussion of *Pithecanthropus erectus* also, slowly but surely, shed new light on Neanderthal Man. He sees Schwalbe's publication on Pithecanthropus as one among many, and thus misses the crucial fact that Schwalbe's Pithecanthropus study led him to his reinterpretation of the Neanderthals.[148] It was precisely Pithecanthropus which widened perspectives to allow a transformist interpretation of Neanderthal Man, because this being obviously fitted into the gap between the ape-man and the modern human.

Erickson suggests, rightly, that Schwalbe's study of Neanderthal Man brought about "a quiet revolution in anthropological thought,"[149] but it was Dubois' discovery of Pithecanthropus that provided the initial impetus for this revolution.[150]

EPILOGUE

We have discussed only a brief episode – 1894 to 1900 – in the debate over the ape-man, but differences of opinion on the interpretation of the fossils continued into the twentieth century. Dubois' material was simply too fragmentary and knowledge of fossil hominids was too incomplete to permit a final verdict on the place occupied by *Pithecanthropus erectus* on the human family tree. Only in the 1930s would Weidenreich's and Von Koenigswald's new discoveries bring in sight a consensus on the genealogical position. We shall be devoting further attention to this in the next chapter.

Besides the lack of sufficient palaeontological material, divergences in ideas on the origin of the hominids were largely responsible for the persisting differences of opinion. To give one example: those researchers who believed that there were human beings as early as the Tertiary, or at all events that Man had reached his present level of development by the beginning of the Pleistocene, relegated Pithecanthropus to a side branch of the human family tree, or regarded him as a relic of a branch that had reached its developmental peak in an earlier era. Marcellin Boule and Arthur Keith were among those adopting this viewpoint.[151] In our discussion of the pre-1900 debate we have already encountered scientists, such as Volz and Hubrecht, who subscribed to a similar opinion.

The debate in the first decades of the twentieth century contributed no essentially new elements. I shall not consider the discussions in this period in detail, because Dubois took very little part in them. Rather, I shall restrict my attention to indicating several aspects of palaeoanthropological research after 1900 which are of interest for our analysis of Dubois' work.

Subsequent discussion of *Pithecanthropus erectus* and other fossil hominids shows that the nineteenth-century anthropological framework of interpretation was being increasingly abandoned. The phylogenetic approach became common property, as did the perception that palaeontological data were required to document the process of humanisation. In whatever way human evolution was conceived, increasing numbers of researchers were coming to recognise that fossils must provide the ultimate solution to the question. We encounter this perception, whether explicitly expressed or just implied, in the work of most of the leading palaeoanthropologists including Schwalbe, Keith, Elliot Smith, Boule, and later on also Gregory, Weinert, Davidson Black, Von Koenigswald, Weidenreich, and Vallois.

At first only a few individuals had both the courage and the means to follow in Dubois' footsteps and actually go in search of fossil hominids. In 1907 and 1908 an expedition, mounted under the auspices of the *Königlich Bayerische Akademie der Wissenschaften* and guided by Margarethe Selenka, went to Java to search for Pithecanthropus fossils near Trinil.[152] The attempt was unsuccessful; although a great number of mammal fossils were collected, no new remains of the ape-man came to light.[153]

The results of the expedition did provide new fuel for the discussion on the age of Pithecanthropus, since the geologists' participating confirmed that the Trinil layers probably dated from the early or even the middle Pleistocene.[154] This raised doubts about whether Pithecanthropus could have been an ancestor of Man at all. According to contemporary understanding the Neanderthals dated from the middle Pleistocene. Did the ape-man and Neanderthal Man live at the same time, or had one form lived in an earlier phase of the middle Pleistocene, the other in a later? Light was only to be shed on this matter decades later, when the structure of the Javanese Pleistocene was better understood.[155]

Roy Chapman Andrews was another scientist to go in conscious search of fossil humans.[156] From 1918 on he led various expeditions to Central Asia, which, according to the American palaeontologist Henry Fairfield Osborn, must have been the cradle of humankind. However, these attempts produced no results either. The important finds at the beginning of this century, such as the discovery of the Heidelberg jaw in 1907, that of the Neanderthal skeletons at La Chapelle-aux-Saints in 1908, and Raymond Dart's discovery of *Australopithecus* in 1924 were accidental finds. Only from the 1930s onwards would deliberately undertaken expeditions produce results and, slowly but surely, become part of normal palaeoanthropological practice.

In the first quarter of the twentieth century there was a huge increase in interest for the research into human descent, also thanks to Dubois' work. The many hundreds of articles on Pithecanthropus appearing over the course of the years bear witness to this interest, but other fossil hominids were also attracting attention. The Neanderthals were studied intensively and Schwalbe's phylogenetic interpretation of this form gradually gained general acceptance.[157] The incredible bru-ha-ha over the putative Piltdown Man would have been inconceivable without the enormous interest the subject had already attracted. We can also see the great attraction of the human-descent question in the many works of a more popular nature published in this period, of which I shall mention only Keith's *Antiquity of Man*, Boule's *Les hommes fossiles*, and Osborn's *Men of the Old Stone Age*. All these books adopted the phylogenetic approach to fossil hominids as their underlying premise, regardless of how the authors believed that human evolution had taken place.

It only remains to discuss what Dubois himself was doing in the field of palaeoanthropology in the first decades of the twentieth century. I shall restrict my comments here to the period 1900–1920, because Dubois again became increasingly active in the 1920s, and we shall be delving more deeply into this matter in the next chapter. However, we must first pause to consider why Dubois withdrew from the Pithecanthropus debate after 1900.

In this post-1900 period Dubois published virtually nothing on Pithecanthropus, at all events nothing new. It is worth mentioning that he gave a lecture in 1911 in which he observed that the development of the ape-man's brain had already progressed so far in a human direction that its place was not exactly halfway between the apes and Man, but rather closer to the latter.[158] He pointed especially to the development of the "lowest frontal brain convolution, so important as the centre of the capacity for speech" in which "the human is already present in a simple form... "[159] This was probably the reason for his declaration, in the same lecture, that Pithecanthropus "must be located close to Man in the system, whether as a separate race or (less probably) within a new family."[160] It is clear that Dubois did not want the family of Pithecanthropidae to be preserved at all costs, as long as Pithecanthropus was recognised as a transitional form.

As far as his ideas on other fossil hominids were concerned, we can assume that Dubois associated himself with the phylogenetic approach, as we can see from several of his statements on the Neanderthals, and the few observations he made on Wadjak Man in this period.

When Dubois first set eyes on the Wadjak fossils during the Sumatran and Javanese excavations, he thought the bones came from a human race and had no direct importance for the question of descent; he attributed only taxonomic significance to the differences between Wadjak Man and the modern human. Yet when the phylogenetic approach came increasingly to be accepted after the turn of the century, he must have realised that his original verdict on the Wadjak fossils required correcting. In 1910 he set himself the task of submitting the fossils to a more rigorous investigation. A letter to C.P. Sluiter dated 17 December 1910 shows the sudden insight he gained by doing this. Dubois told Sluiter that he had begun to prepare the

Wadjak fossils – until then he had not cleaned off the sediment still clinging to the specimens – and reported:

> And it is now obvious that everything is so important that I feel the need to share it with you, without delay. I had... recognised earlier on that I was dealing with an Australian type, not with a Papuan; now closer acquaintance with the jaws, particularly, shows me that this [type] is of a very primitive kind and resembles the neanderthaloids in many respects. Actually it is good that the task has been left in abeyance for so long, as the Australians and [the Neanderthals] have only recently become better known and [since] our early Javanese is in the line, which on the basis of that knowledge is believed to continue further back into the past.

Dubois now located the Wadjak in an evolutionist context and interpreted the morphological differences from the modern human in a phylogenetic sense.

The letter to Sluiter also suggests that he had revised his opinion on the Neanderthals. This is confirmed by the following quotation taken from the 1911 lecture:

> Where the fossils of Homo primigenius are concerned there can be no doubt that a human form lived in the diluvial era in Europe, standing on a lower rung of the developmental ladder than present-day Australians and all other low-level races... Since the remains of Homo primigenius... have become better known we have ceased to doubt that this extinct human form is indeed somewhat closer to the anthropoids than the extant species of Man.[161]

Thus in the early decades of the twentieth century Dubois also switched to the phylogenetic interpretation of the Neanderthals propagated by Schwalbe, who, for that matter, suggested the name *Homo primigenius* which Dubois used in the quotation given above.

Finally, why did Dubois withdraw from the discussion and for something like twenty years refuse to allow anyone to see the fossils? John Reader's explanation in his book *Missing Links*, can serve as a model for the reasons usually given in the literature on Dubois:

> [Dubois'] conclusion invited comment and scientists travelled to Holland to examine the fossils and discuss their interpretation. But Dubois found these visits increasingly tiresome, especially as so few of the visitors shared his views. Science had honoured him with gold medals, diplomas and honorary degrees in recognition of his work, but it would not give him

what he most wanted – general agreement with his belief that *Pithecanthropus erectus* was neither ape nor man but a link between the two. Finally Dubois retaliated by severely restricting access to the fossils.[162]

We can indeed conclude from Dubois' correspondence that he had an infinite capacity for annoyance with any scientist whose viewpoint differed from his own, and he was also very ready to ascribe this to contrariness. His suspicious nature, verging on the paranoid, sometimes caused him to suspect dark motives behind colleagues' disagreement with his views. The older he grew the more convinced he became, for instance, that 'the Catholics' were out to destroy the fossils.[163] Reader's explanation is therefore plausible in every respect.

However, I think something else was also operating here. Reader does not explain why Dubois refused access to the fossils to supporters as much as opponents. It is also intriguing that Dubois withdrew just at the time when his following was increasing; as we saw, more and more researchers were accepting his viewpoint in the last years of the nineteenth century. An additional explanation, taking these factors into account, might be as follows.

Dubois came back to the Netherlands to publish a comprehensive work on *Pithecanthropus erectus* and the Javanese fossil fauna, a work that was to establish his reputation. The professorship in Amsterdam he accepted in 1898 was not a particularly prestigious one, and therefore Pithecanthropus and the collection of fossil vertebrates constituted his capital for investment in a career, so to speak. This encourages the suspicion that Dubois decided to place an embargo on viewing the fossils partly because he quite simply did not want others stealing his thunder. He had behaved generously to Schwalbe in 1897, giving him the opportunity of viewing the Pithecanthropus fossils and sending him a cast of the skullcap. Schwalbe thereupon produced a major publication on the cranium which far overshadowed Dubois' description. Dubois must have realised that, if he continued in this way, there would soon be little credit to gain by producing a new description.[164] He probably drew a lesson from this affair and decided to keep the fossils to himself until he had written about them.

The scientific world had to wait years for this description, however, and Dubois himself never produced a full treatment of the vertebrate fossils. Because of this outsiders were for years on end denied access to the collection, until the Royal Dutch Academy of Science finally intervened, in the 1920s, and more or less compelled Dubois to agree to a compromise.[165]

What still requires explanation is why the description of the vertebrate fossils never came to anything of much note. Work on the collection was to dominate Dubois' life until long after his retirement, not because he was always actively involved in it but because he never got around to it, despite all the plans, promises, and pressure from outside. Until the 1930s he also continued to demand that he and he alone should write the description, and he mounted a tooth-and-nail resistance to any outside interference.[166] Apart from a monograph on *Manis palaejavanica*, the giant pangolin of Java, all that he published on the collection was a couple of articles, written in 1907 and 1908, which were little more than congeries of species names and general conclusions on the age of the Kendeng fossils.[167] The reason for publishing these articles fits the pattern we have outlined: the Selenka expedition was at that time working at Trinil, and Dubois wanted to ensure his priority.

To find a satisfactory explanation for Dubois' curious behaviour, a deeper biographical investigation than mine is needed. I can only point to a feature of his personality that may have been partly responsible for this, by quoting (for the second time) a statement by one of Dubois' students: "Dubois had the habit of just lifting a corner of the veil of a scientific conception, but he was loath to settle down and work it out thoroughly."[168] It could well be that the prospect of describing the fossil collection simply bored him; perhaps he had neither the peace nor the patience needed for the task. Also, he found field work more attractive, and in the period we have been examining he always found time for it, since many of his studies on the geology and palaeontology of the Netherlands were undertaken in the first two decades of this century.

Notes

1 For the period up to and including 1899 Quenstedt mentions, in his bibliography of palaeoanthropological writings, 78 articles and books by 47 authors in which Dubois' find is discussed (*Hominidae Fossiles*, 175–177). I have made a selection from these publications, founded on the criteria mentioned, which in my view gives a representative picture of the reception given to *Pithecanthropus erectus*. I have been unable to trace five of the publications cited by Quenstedt, but conversely I have discovered a number of publications that are not mentioned in his bibliography. I discuss here 54 reactions from 41 authors.

2 Anon., "Belangrijke palaeontologische vondsten".

3 *Ibid.*, 312.

4 Matschie, "Anthropopithecus erectus".

5 Manouvrier, "Discussion du 'Pithecanthropus erectus'". A summary of this article appeared under the title "Le Pithecanthropus".

6 *Ibid.*, 40.

7 Pettit, "Pithecanthropus erectus".

8 Krause, Discussion of Dubois' *Pithecanthropus erectus.*

9 *Ibid.*, 80.

10 *Ibid.*, 79.

11 Von Luschan, Commentary on Krause's discussion of Dubois' *Pithecanthropus erectus.*

12 Waldeyer, Commentary on Krause's discussion of Dubois' *Pithecanthropus erectus.*

13 Virchow, Commentary on Krause's discussion of Dubois' *Pithecanthropus erectus.*

14 L[ydekker], Review of Dubois' *Pithecanthropus erectus.*

[15] Cunningham, "Dr. Dubois' So-Called Missing Link".
[16] *Ibid.*, 428.
[17] *Ibid.*, 428–429.
[18] *Ibid.*, 429.
[19] Cunningham, [A Paper on Pithecanthropus erectus].
[20] Keith, Commentary on Cunningham's [A Paper on Pithecanthropus erectus].
[21] Keith, "Pithecanthropus erectus".
[22] *Ibid.*, 368.
[23] *Ibid.*, 369.
[24] Turner, "On M. Dubois' Description".
[25] *Ibid.*, 445.
[26] Topinard, Review of Turner's "On M. Dubois' Description".
[27] *Ibid.*, 607. The American palaeontologist E.D. Cope also shared the opinion that Dubois had demonstrated the presence of Neanderthals on Java ("The Neanderthal Man in Java", *The American Naturalist* 29 (1895) 192–193; non vidi).
[28] Marsh, "On the Pithecanthropus erectus, Dubois".
[29] *Ibid.*, 147.
[30] Matschie, "Noch einmal Anthropopithecus erectus".
[31] Ten Kate, Review of Dubois' *Pithecanthropus erectus.*
[32] Martin, "Kritische Bedenken".
[33] *Ibid.*, 216.
[34] *Ibid.*, 217.
[35] Virchow, [Die Frage von dem Pithecanthropus].
[36] Haeckel, *Systematische Phylogenie*, 617, 631, 633–634. Besides striving for recognition of Dubois' work in his writings, Haeckel also made more practical efforts. In 1895 he suggested to the faculty of philosophy at Jena University that they make Dubois an Honorary Doctor. However, the proposal met with a great deal of opposition, and was turned down. Correspondence between Dubois and Haeckel on this matter can be found in the Dubois Collection.
[37] *Ibid.*, 617.
[38] *Idem.*
[39] *Ibid.*, 616.
[40] *Ibid.*, 618.
[41] Dubois, "Pithecanthropus erectus, eine menschenähnliche Uebergangsform", 1896. This lecture also appeared, in a somewhat different form, under the title "Näheres über den Pithecanthropus".
[42] *Ibid.*, 251–257.
[43] *Idem.*
[44] *Ibid.*, 258–260. In a letter Dubois had asked Léonce Manouvrier to see if he could find thigh bones resembling that of Pithecanthropus in the osteological collections in Paris.
[45] *Ibid.*, 256, 260–261.
[46] *Ibid.*, 261–262.
[47] *Ibid.*, 262–264.
[48] *Ibid.*, 264–265.
[49] *Ibid.*, 265–267.
[50] *Ibid.*, 268–270. Compare [Dubois], "Palaeontologische onderzoekingen" (4th quarter 1892), 1893, 11.
[51] Dubois, "Pithecanthropus erectus, eine menschenähnliche Uebergangsform", 1896, 268–270.
[52] *Ibid.*, 267–269.
[53] *Ibid.*, 270.

[54] Marsh, Rosenberg and Martin, Commentary on Dubois' "Pithecanthropus erectus, eine menschenähnliche Uebergangsform". Compare also Krause, "Der III. internationale Zoologencongress", 540–541.
[55] See Anon., "The Third International Zoological Congress", 554; Krause, "Der III. internationale Zoologencongress", 541; [J.F.] v[an] B[emmelen], "Dubois' aapmensch".
[56] Krause, "Der III. internationale Zoologencongress", 541.
[57] J.M. van Bemmelen, "Der Gehalt an Fluorcalcium"; "Sur la teneur en fluorure de calcium".
[58] Dubois, "Le Pithecanthropus erectus du pliocène de Java", 1895.
[59] Dubois, "On Pithecanthropus erectus: a Transitional Form", 1896, 13. A summary of this lecture appeared in the *Journal of the Anthropological Institute.*
[60] *Idem.*
[61] Dubois, "Pithecanthropus erectus betrachtet als eine wirkliche Uebergangsform", 1895.
[62] *Ibid.*, 727–730.
[63] *Ibid.*, 731–733.
[64] *Ibid.*, 735–736.
[65] *Ibid.*, 737.
[66] Dubois, "De thans bekende fossiele menschapen", 1896, LXX–LXXII. See also Dubois, "Ueber drei ausgestorbene Menschenaffen", 1897, 84–91. It is clear from the phylogenetic tree that Dubois exhibited in Berlin (see Fig. 11) that he was already having doubts about the kinship between *Anthropopithecus sivalensis* and the chimpanzee; at all events he switched to the name *Palaeopithecus.* At that point he obviously still believed that *Palaeopithecus* could have been the ancestor of Pithecanthropus. In the articles mentioned above Dubois also discussed two other fossil primates, *Pliopithecus antiquus*, dicovered by Lartet (see chapter 2, note 62) and a thigh bone attributed to *Dryopithecus*, discovered in 1820 in the vicinity of Eppelsheim. In Dubois' view, earlier authors had been incorrect in their assumption that there was a close relationship between *Pliopithecus* and the gibbons. In contrast the fossil from Eppelsheim did resemble the gibbons in some respects, which militated against its inclusion in the genus *Dryopithecus.* Accordingly Dubois proposed a new name for the latter: *Pliohylobates eppelsheimensis* (see also Dubois, "Le Pithecanthropus erectus du pliocène de Java", 1895, 154–155). Dubois became involved in a controversy with Max Schlosser over his interpretation of the Eppelsheim thigh bone (see Dubois, "Zur systematischen Stellung", 1901; see the same for further references). The discussion with Schlosser is not relevant for us.
[67] Dubois, "Le 'Pithecanthropus erectus' et l'origine de l'homme", 1896, 467.
[68] Dubois, "The Brain-Cast of Pithecanthropus", 1899, 96.
[69] Dubois, "Pithecanthropus erectus, eine Stammform des Menschen", 1896.
[70] *Ibid.*, 8–9.
[71] *Ibid.*, 15–16.
[72] Dubois, "The Brain-Cast of Pithecanthropus", 1899.
[73] *Ibid.*, 81. After Dubois had left, the engineer Kriele continued the excavation work up to and including 1900, with the exception of 1898, in which year no digging was carried out. Kriele concentrated on Trinil and on localities in the Solo valley. At regular intervals he sent Dubois any fossils discovered. See De Vos and Sondaar, "The Importance of the 'Dubois-Collection'", 42–43.
[74] Dubois, "The Brain-Cast of Pithecanthropus", 82–85.
[75] Dubois, "Données justificatives", 1902.
[76] *Ibid.*, 237–239.
[77] *Ibid.*, 239–241. See also Dubois, "The Brain-Cast of Pithecanthropus", 1899, 80.
[78] *Ibid.*, 241.
[79] Manouvrier, "Deuxième étude", 555. A summary of this article appeared under the title "Le Pithecanthropus erectus et la théorie transformiste".

[80] *Ibid.*, 628–629.
[81] *Ibid.*, 645.
[82] *Ibid.*, 648.
[83] De Mortillet, "Précurseur de l'homme". Compare also his "l'Anthropopithèque".
[84] *Ibid.*, 316.
[85] According to Verneau, "Encore le Pithecanthropus erectus", 726.
[86] *Idem.*
[87] Marsh, "On the Pithecanthropus erectus". An article along similar lines appeared under the title "The Ape-Man from the Tertiary of Java".
[88] *Ibid.*, 478.
[89] *Ibid.*, 482.
[90] Nehring, Commentary on Dubois' "Pithecanthropus erectus betrachtet als eine wirkliche Uebergangsform". See also his "Menschenreste".
[91] Dames, "Pithecanthropus". Jaekel, Commentary on Dubois' "Pithecanthropus erectus betrachtet als eine wirkliche Uebergangsform".
[92] Haeckel, "Unsere gegenwärtige Kenntniss".
[93] *Ibid.*, 26–27.
[94] Virchow, [Weitere Mittheilungen]; "Der Pithecanthropus vor dem zoologischen Kongress"; Commentary on Dubois' "Pithecanthropus erectus betrachtet als eine wirkliche Uebergangsform".
[95] Virchow, Commentary on Dubois' "Pithecanthropus erectus betrachtet als eine wirkliche Uebergangsform", 744.
[96] Virchow, "Exostosen und Hyperostosen",
[97] Virchow, Commentary on Dubois' "Pithecanthropus erectus betrachtet als eine wirkliche Uebergangsform", 748–749.
[98] Volz, "Pithecanthropus erectus"; "Demonstration eines Schädel-Abgusses". Bumüller, "Menschen- und Affen-Femur".
[99] Hubrecht, "Aap of mensch?", 518.
[100] Branco, "Die menschenähnliche Zähne", 98–112. The quotation is taken from p. 100.
[101] Kollmann, Commentary on Dubois' "Pithecanthropus erectus betrachtet als eine wirkliche Uebergangsform".
[102] Klaatsch, "Der gegenwärtige Stand"; "Die Stellung des Menschen".
[103] Martin, "Weitere Bemerkungen".
[104] Houzé, "Pithecanthropus erectus".
[105] *Ibid.*, 8.
[106] *Ibid.*, 27.
[107] *Ibid.*, 36.
[108] Manouvrier, "Réponse aux objections".
[109] *Ibid.*, 453.
[110] *Ibid.*, 451.
[111] Anon., Report on a meeting at the Anthropological Institute, 249–251.
[112] *Ibid.*, 250.
[113] *Idem.*
[114] *Idem.*
[115] Hepburn, "The Trinil Femur".
[116] Anon., Report on a meeting at the Anthropological Institute, 248.
[117] *Ibid.*, 253–254; 254–255.
[118] *Ibid.*, 252–253.
[119] Anon., "Dr. Dubois' 'Missing Link'", 115.
[120] Cunningham, "The Place of 'Pithecanthropus'". This paper was published in response to Dubois' incorrect reporting (in his "The Place of 'Pithecanthropus'", 1895–1896) of

Cunningham's ideas on the phylogeny of Pithecanthropus. Dubois in his turn had reacted in his article to what seemed to him to be a misleading exposition of his ideas in Anon., "Dr. Dubois' 'Missing Link'".

[121] Anon., Report on a meeting at the Anthropological Institute, 251–252.

[122] Anon., "Dr. Dubois' 'Missing Link'", 116.

[123] Sollas, "Pithecanthropus erectus". See also Anon., "Dr. Dubois' 'Missing Link'", 116.

[124] Newton, "Palaeolithic Man". Anon., Report on a meeting at the Anthropological Institute, 249.

[125] Munro, "On Intermediary Links".

[126] *Ibid.*, 250.

[127] Schwalbe, "Studien über Pithecanthropus". For Schwalbe, see Keibel, "Gustav Albert Schwalbe"; Fischer, "Gustav Schwalbe".

[128] *Ibid.*, 16–17.

[129] Schwalbe, "Studien über das Femur von Pithecanthropus".

[130] Schwalbe, "Studien über Pithecanthropus", 222.

[131] *Ibid.*, 225–226.

[132] *Ibid.*, 228.

[133] Schwalbe, "Der Neanderthalschädel".

[134] See for example Schwalbe, "Vorgeschichte des Menschen". Schwalbe also acknowledged the possibility of parallel evolution of the three forms from common Tertiary ancestors.

[135] For this, see Jacob, "Virchows Begriff der naturwissenschaftlichen Methode"; Ackerknecht, *Rudolf Virchow*, 165–172.

[136] Ackerknecht, *Rudolf Virchow*, 167–168.

[137] According to Bumüller in his "Menschen- und Affen-Femur", 91.

[138] For Ranke, see Rückert, "Johannes Ranke".

[139] See for example Day and Molleson, "The Trinil Femora".

[140] Compare chapter 1, p. 7.

[141] Lodge, "Turner, Sir William", 539.

[142] Schwalbe, "Ziele und Wege einer vergleichenden physischen Anthropologie", 2.

[143] *Ibid.*, 5.

[144] *Ibid.*, 15.

[145] Erickson, *Origins of Physical Anthropology*, 200–206, 215–230.

[146] *Ibid.*, 222.

[147] *Ibid.*, 220, 224.

[148] *Ibid.*, 222.

[149] *Ibid.*, 205.

[150] The fact that, also thanks to Dubois' and Schwalbe's work, palaeoanthropology was able to break away from nineteenth-century ideas on race, naturally does not imply that their work was solely responsible for the dismantling of the anthropological framework of interpretation of fossil hominids. The criticism of the theories and methods on which this framework was based derived mainly from research in other disciplines. The Aryan thesis – at all events the anthropological version of this – had increasingly fallen into discredit since 1890. The decline of the theory was set in motion by criticism from ethnologists and philologists as much as by anthropologists (see Erickson, *Origins of Physical Anthropology*, 113–114; Poliakov, *The Aryan Myth*, 256, 327). The monogenism-polygenism debate, with all its theological and political overtones, was burnt out by the end of the nineteenth century, even though the scientific problem remained unsolved. Polygenist theories in which human races had derived from different species of apes remained in vogue for some time (see for example Klaatsch, "Die Aurignac-Rasse", 567). The phylogenetic interpretation of fossil hominids did nothing at first to clear up the difficulties, since extinct hominids could be fitted into both polygenist and monogenist conceptions of human evolution. The reduction-

ist and typological methods within research on the human races continued to be used until after 1900. Ethnographers, especially, emphasised the need for investigation of the 'nature' side of the 'nature-nurture' spectrum. Franz Boas took important initiatives in this direction after the turn of the century (see for example his "The Aims of Anthropological Research"). The static and typological view of the human races was finally replaced by a more dynamic one only when research into genetics – especially population genetics – began to be carried out in the first decades of this century (see for example Penniman, *Hundred Years*, 182–190).

151 See for example Boule, *Les hommes fossiles*, chapter 4; Keith, *The Antiquity of Man*, vol. 2, chapter 22. The theories of human evolution in the period 1844–1944 have recently been examined by P.J. Bowler, in his *Theories of Human Evolution*. This book came to my attention too late to be utilised in this study.

152 The results of this expedition were published in Blanckenhorn, *Die Pithecanthropus-Schichten*.

153 The only discovery of (subfossil) hominid remains was that of a fully human tooth, uncovered at Sondé, some distance from Trinil. However, the age of the specimen was uncertain. Dubois joined in the discussion arising from this find and – naturally – refused to accept that the tooth could possibly have come from the same layer as the Pithecanthropus remains. He even suggested that there was fraud somewhere. See Dubois, "Das geologische Alter", 1908, 1252 note; Selenka, [Die fossile Zähne von Trinil]; Dubois, [Die fossile Zähne von Trinil], 1909; Selenka, "Der menschliche Zahn"; Walkhoff, "Der Zahn von Sondé".

154 For this, see for example Volz, "Das geologische Alter"; Elbert, "De nieuwste onderzoekingen"; Branco, "Vorläufiger Bericht", 266–273; Schuster, "Die Flora der Trinil-Schichten".

155 See Oakley, "The Problem of Man's Antiquity", 133–143.

156 See for example his *On the Trail of Ancient Man*.

157 See Erickson, *Origins of Physical Anthropology*, 202–206.

158 Dubois, "Over de plaats van Pithecanthropus", 1912, 4–8.

159 *Ibid.*, 6.

160 *Ibid.*, 7.

161 *Ibid.*, 5.

162 Reader, *Missing Links*, 49.

163 See for example Idenburg, *De Leidse universiteit*, 38–40.

164 From the quotation given below, taken from Schwalbe's letter to Dubois on 20 December 1897, we can see that Dubois made clear his concern about this to Schwalbe. The latter's letter certainly would not have laid his fears to rest: "I shall be glad to leave the femur to you for some time. Certainly we are competitors... I have almost finished a manuscript on the skull and another one on the femur... We will both have to accept [the fact that] one or the other will finish sooner."

165 For more on this, see chapter 5.

166 There is a voluminous correspondence in the Dubois Collection on this affair. The continual postponement of Dubois' description even led to questions in parliament; for this see Tillema, *Victor de Stuers*, 216–219. For the fate of the collection between 1895 and 1940, see Brongersma, "De verzameling van Indische fossielen".

167 Dubois, "Eenige van Nederlandschen kant verkregen uitkomsten", 1907; "Das geologische Alter der Kendeng-Fauna", 1908; "Manis palaejavanica", 1926.

168 Van der Feen and Van Benthem-Jutting, "Dr. Antje Schreuder", 122.

Chapter 5

CEPHALISATION, PITHECANTHROPUS, AND EVOLUTION

Introduction

From 1897 on Dubois was occupied with research on cephalisation,[1] a subject arising directly from his study of the Pithecanthropus fossils. He began by working on this at irregular intervals, but the research gained momentum after 1918.

In the course of the 1920s Dubois' palaeoanthropological work began to dovetail into his research on cephalisation, and his ideas on human descent increasingly derived from the results of his investigation of the brain. Together with this development, Dubois' view on the evolutionary process and its underlying mechanisms changed radically. Step by step he formulated a model of evolution which was completely at odds with that of Darwin; its tenets were saltatory evolution and the presence of an autonomous developmental force. By 1930 he had achieved a definitive formulation of this new model.

For a proper understanding of Dubois' later publications on Pithecanthropus it is essential to recognise the interconnection between his cephalisation research, his palaeoanthropological studies, and his view of the evolutionary process. In much of the existing literature on Dubois this interlinking is not recognised, resulting in a distorted image of his ideas.

In this chapter I shall first provide an outline of Dubois' studies of the brain and of the development of his conception of the evolutionary process. Taking this as our background we shall then examine Dubois' palaeoanthropological investigations in the period 1920–1940, stressing his publications on Pithecanthropus. Next I shall give an account of his reaction to the discoveries of Sinanthropus and Pithecanthropus in the 1930s, and of his opposition to Weidenreich's and Von Koenigswald's interpretation of these fossils. The historical context and significance of Dubois' research into cephalisation will be discussed in the final section.

THE THEORY OF CEPHALISATION

We have seen how, in the debate over interpretation of Pithecanthropus, Dubois argued on several occasions that the fossils could not have derived from an ape since in that case the body weight would have had to be several hundred kilos, which conflicted with the dimensions of the femur.[2] This argument was based on the assumption that within each group of animals there is a constant ratio (determined by their place within the natural system) between brain weight and body weight. In 1897 he tried to provide empirical support for this thesis and to couch it in a mathematical form, and he also attempted to produce a biological explanation for the supposed link between brain and body weights. By a happy coincidence the Amsterdam zoologist Max Weber had published, in 1896, a survey of the brain and body weights of a large number of mammal species.[3] This presented Dubois with an ideal springboard for his analyses.

Dubois began his first article on the subject, published in 1897, with a theoretical analysis of the relationships under investigation.[4] He believed that two factors determined an animal's brain weight: firstly, the size of the body represented by bodily weight, and secondly, the developmental level of the central nervous system, which he specified as the degree of cephalisation. Although the nature of the relationship was still unclear, in his view there had to be a link between brain weight and body weight. Experience had shown that small animals possessed relatively bigger brains than large animals, a phenomenon already recognised by Cuvier. Alexander Brandt, Theodor Bischoff, Max Fürbringer, and Otto Snell had suggested that the higher metabolic rate among small animals, caused by their relatively larger body surface, was directly responsible for their comparatively bigger brains. These authors believed that the brain regulated the metabolism, and Fürbringer had even posited the presence of an extensive heat centre in smaller animals' brains.[5]

Assuming that body surface reflected the intensity of the metabolism and thus also the weight of the brain, Snell had formulated the link between brain and body weights in the following manner. If P, the body weight, is taken as a measure of body volume then body surface can be written as $P^{2/3}$ or $P^{0.66}$. Thus in two closely related animals (a cat and a lion, for instance) with weights P_1 and P_2, differing only in size and not in the level of organisation within the central nervous system, the brain weights have the ratio $P_1^{0.66} : P_2^{0.66}$. In comparing animals with differing levels of organisation the value $P^{0.66}$ has to be multiplied by a factor c, the cephalisation coefficient, representing the degree of brain development.[6] In this case the ratio of the brain weights is thus $c_1P_1^{0.66} : c_2P_2^{0.66}$.

Dubois agreed with this mathematical formulation but he felt that the connection had a completely different background. Snell had erroneously made a direct connection between two phenomena that coexist side by side and have an indirect influence on each other. With all that is now known about the structure and function of the brain, these [phenomena] are only to an extremely limited extent what Brandt and Snell and especially Fürbringer claim they are, an organ that regulates the metabolism.[7]

Dubois believed that the structure and action of the nervous system, rather than metabolism, was responsible for the supposed mathematical connection. Basing his case on the neuron theory worked out in the second half of the nineteenth century by Wilhelm His, August Forel and others, Dubois described the nervous system as consisting "entirely of connecting arcs between sensory and motor nerves, the complexity of which is greater or lesser according to the animal's higher or lower organisation."[8] Primary arcs, or sensory-motor units, linked the senses to the muscle system via the spinal cord. Secondary arcs created the link with the brain, and the complexity of these arcs in the brain determined the developmental level of the central nervous system. Dubois reasoned that in mammals with the same level of organisation but of different size, the degree of complexity in the nervous systems was identical, but there was a difference in the total perceptive surface of the sense organs which enabled the animals to maintain contact with their surroundings. This implied that different numbers of sensory-motor units were involved. When body surface was used as a measure of the perceptive area of the senses, then brain weight was in truth connected with body surface in accordance with Snell's formula, but for reasons different from those which Snell had posited.[9]

This was naturally a very simplistic picture, Dubois continued. Various other factors had their role in the matter, and could upset the simple mathematical relationships. He named two of these factors. Firstly, the cerebral cortex – the most important part of the brain, functionally speaking – was approximately of equal thickness in all mammals, great or small. The reason for this was the need for a continuous blood supply in the cortex, which was provided via the cortex surface. Consequently, for the increase in cortex volume needed by larger animals, the total brain weight had to increase faster than as $P^{0.66}$, or the cortex had to become convoluted. Dubois suggested that both these things often occurred at the same time, complicating the picture considerably.[10]

A second factor disturbing the simple ratios derived from the fact that small animals were relatively more strongly innervated than large ones, as was apparent from their faster movements, for instance. The fact that eye size in smaller animals was larger than one would expect from their surface measurements, pointed in the same direction. The number of sense elements per surface unit probably increased more than expected; therefore the number of sensory-motor units also increased, and hence brain weight as well.[11] However, Dubois believed that these factors could in their turn be assumed to exert an influence, corresponding to body size. There still had to be a fixed relationship between brain and body weight, even though this might not be determined by precisely $P^{0.66}$.[12]

How did this model work out in practice? Dubois felt that "the validity of these inferences can be proved in a truly striking manner from our present knowledge of the central nervous system... "[13] Taking the animals Weber had weighed, Dubois grouped them in pairs of closely related species with an assumed equality in the degree of cephalisation. When the brain and body weights of these pairs were fitted into Snell's formula, the resulting values of the exponent were approximately 0.56. Thus the values were in the order of magnitude of 0.66, while the difference from this value was more or less constant. According to Dubois the fact that the exponent was smaller than 0.66 had to be ascribed to the fact that small animals were more strongly innervated. At all events he felt that there could be no doubt about the accuracy of the explanatory model: "Thus it is actually the size of the perceptive surface of the sense organs that determines quantity in brains of equally high organisation."[14]

Taking the value 0.56 as his starting point, Dubois now calculated the c-values or cephalisation coefficients of animals unequal in organisational level with the formula $E = cP^{0.56}$, with E representing the brain weight. The coefficients obtained by this method, providing a relative measure for the degree of organisation, clearly accorded

> much better than other [measures] for the relative quantity of the brains in different animals, with the ranking order occupied within the [natural] system by the animals concerned, and also with what we know about their 'mental capacities'.[15]

The human occupied the first position in the series of cephalisation coefficients, followed in descending order by the apes, the hoofed mammals and carnivores, the rodents, the insectivores, and so on.

Finally Dubois warned against overestimating the accuracy of the values he had found. In the first place, because data were not available for every animal species, the average value of the exponent was taken as a starting point, and it was clear that this value deviated considerably from 0.56 in some groups of animals. In the second place one had to bear in mind that the level of organisation was not determined solely by the quantity of the brain, but also by its quality.[16]

These qualifications apart, Dubois considered that the belief in the existence of a systematic link between brain and body weights was completely justified. He thought that he now had a means of finding out the degree of organisation in *Pithecanthropus erectus*, and at the Fourth International Congress of Zoology at Cambridge in 1898 he presented his results.[17]

Firstly, the brain weight of Pithecanthropus had to be ascertained. Dubois fixed the contents of the skullcap – by then completely cleaned out – at 570 cc, but since the fossil was incomplete the total volume must have been larger. When measured according to the ratio between skullcap volume and total volume among humans, the capacity must have been larger by a factor of 1.35–1.45, and measured by the same ratio among ape skulls by a factor of 1.43–1.58. Dubois took 1.5 as his factor and arrived at the figure of 855 cc – 50 cc lower than in his calculations published in 1896. From this a brain weight of approximately 750 grams could be inferred.[18]

By means of a series of calculations Dubois then demonstrated that the cephalisation coefficients in modern humans of all races had roughly the same value; at most the variation was 10%. In the event that Pithecanthropus had reached the same degree of cephalisation as the modern human, its brain weight of 750 grams would have given it a body weight of 19 kilos, an absurd value. If one began with the cephalisation coefficient for the great apes, the resulting body weight would have been 230 kilos, which was equally unacceptable; the dimensions and supporting power of the fossil femur were

Fig. 15. Dubois and several other participants of the zoological conference at Cambridge in 1898.

not consistent with a weight of this kind. The thigh bone indicated a body weight of 70–75 kilos, and on this basis one could calculate that the fossil's cephalisation coefficient was roughly half that of the human and double that of the apes. The mathematical relationship between brain and body weight thus confirmed the intermediate position of Pithecanthropus in a manner as simple as it was effective.[19] Consequently Dubois concluded that: "From all these considerations it follows that Pithecanthropus erectus undoubtedly is an intermediate form between Man and the Apes."[20]

With this, Dubois had achieved the intended purpose of his studies into brain and body weight, and his research into cephalisation was brought temporarily to a close.[21] Yet he returned to the subject at the end of 1913, this time leaving Pithecanthropus out of the picture.[22] The reason for returning to the question was probably a number of articles by the French physiologist Louis Lapicque, who had become engrossed in the subject after Dubois' first publications appeared. Lapicque's results offered Dubois the chance to extend his theory still further.

By and large Lapicque had been able to confirm Dubois' findings. The exponent 0.56 which Dubois had established for mammals was seen to be applicable to birds as well, and the classification of birds according to their cephalisation coefficients corresponded reasonably closely to the kinship relations among birds within the natural system.[23] Dubois was now able to provide additional data on reptiles, amphibians, and fishes, indicating that an exponent of roughly 0.56 was valid for these creatures as well. Thus the same exponent of approximately $^5/_9$ could be applied to the entire group of vertebrates, and Dubois concluded that a law could be perceived which he formulated in the following way:

> *In species of Vertebrates that are equal in organisation (systematically), in their modus of living and in shape, the weights of the brains are proportional to the $^5/_9$ power of the weights of the bodies.*[24]

With the help of Snell's formula Dubois subsequently worked out the relation between brain and body weight for individuals of the same species. Lapicque had also been occupied with this problem, and both he and Dubois had achieved similar results, Dubois finding an exponent of roughly 0.25 for the human, Lapicque 0.224 for women, 0.23 for men, and 0.25 for dogs.[25] The same calculation, applied to several species of mammals, gave Dubois a figure of approximately 0.23. For him it could safely be assumed that the same inter-individual exponent of approximately $^2/_9$ was valid for all vertebrates, and that here also, one could see a scientific law in operation.[26]

In the second part of his article Dubois tried to discover a biological explanation for the values he had discovered. Lapicque believed that the exponents constituted only the empirical resultants of unknown but very complex interconnections.[27] In contrast Dubois was firmly convinced that the exponents possessed an immediate biological significance, the mathematical relations representing biological laws, and he even thought he could find proof for this. Here I shall reproduce Dubois' argument in concise form, as an example of his way of reasoning.

His starting point was that the number of sensory-motor units in the body could serve as a measure of brain weight. Firstly, the terms $P^{2/9}$ and $P^{5/9}$ ($= P^{2/9 + 3/9}$) could be written as $L^{2/3}$ and $L^{2/3} \times L$ respectively, L representing the linear dimension of the body, so $L = P^{1/3}$. Thus both terms included $L^{2/3}$ or $P^{2/9}$ as their common factor. In Dubois' view the biological meaning of this factor could be deduced from the construction and function of the eye.

The eye functioned as an organ for spatial perception, its most important task being to direct movements. Therefore eye size, Dubois reasoned, must correlate with all the nerve paths connected with this function; in other words, there had to be a link between, on the one hand, the energy conversion in the eye resulting from stimulation of the retina, and on the other hand the motor energy generated as a response. Now if the body length increased as L, the muscle power increased correspondingly as L^2 (that is, according to the increase in the cross section of the muscles) and body mass as L^3. Body mass thus increased faster than muscle power by a factor L and it followed that the larger animal needed an L-fold stimulus to move its bulk. If the mass of the sense cells was a measure of their power of energy conversion, then this mass in the retina must increase as L. The linear increment was thus as $L^{1/3}$, while that of the surface was as $L^{2/3}$ or $P^{2/9}$. If at the same time this surface increase corresponded to the growth in the number of optical cells in the retina – and thus also to the rise in the number of all the sensory-motor units correlated with the eye – then the brain mass must also increase as $P^{2/9}$. This explained the factor $P^{2/9}$, which Dubois dubbed the 'eye factor'.

The remaining factor, $P^{3/9}$ or L, represented for Dubois an increase in the number of nerves in relation to the increase in length from species to species. One could not expect anything else among metameric (segmented) animals, he thought, and consequently called the term $P^{3/9}$ the 'segmental factor'.

In Dubois' view this explanation was supported by measurements of eye diameter among related animal species of varying size. According to measurements taken by Lapicque, among others,[28] the diameter of the eye, which could be considered a measure of the retina diameter, varied with brain size in the same way as body weight did: in relation to body weight the eye diameter increased as $P^{1/7.5}$, which was a very small deviation from the expected value of $P^{1/9}$ ($L^{1/3}$). Dubois thought the deviation could be explained by the (theoretically necessary) increase in the thickness of the eye nerves because of which the sensory surface also grew.[29]

This 'biological explanation', ingenious as it may seem, was of course an informed guess at best, since Dubois' reasoning depended on the accuracy of the rather large number of hypotheses on the construction and working of the nervous system. It seems that even Dubois himself did not regard the accuracy of his assumptions as being beyond all doubt. In a German version of the article appearing later on, he gave the impression of doubting whether it was actually the number of sensory-motor units that decided brain size, or rather their total mass. His statements on this point were ambiguous at all events. He continued to maintain that the exponents had a biological meaning, and retained the notion of eye factor and segmental factor, but he left part of his biological explanation out of the German version.[30]

In the last part of his article Dubois argued that the suggested explanatory model for the link between brain and body weights made it possible to produce a functional explanation of the differences in cephalisation among related animals. Given that brain size was connected with the perceptive surface of the sense organs, there also had to be a direct link with animal functions and with their degree of specialisation. Here I shall give some of the examples Dubois used to illustrate this point:

> According to numerous evidences the cephalisation of the Brown Rat and the Black Rat and likewise that of the Housemouse is half that of Hares (and Rabbits) and only a third part of that of Squirrels. In the Hares the sense of hearing, in the Squirrels... the organ of touch, on account of its high specification (in the hand), has caused the increase of the brain... The Elephant surpasses the other Hoofed Mammals three times in cephalisation. He ranks even much higher than the Anthropoid Apes. He owes this to his trunk, which has become a prehensile and touch hand... Some of the American Monkeys (Ateles), which are higher cephalised than the Monkeys of the Old World, *not excepted the Anthropoid Apes*, obtained a third prehensile and touch hand in their tail.[31]

The biological meaning of the exponents continued to occupy Dubois' attention. When he returned to the subject in 1918 he approached the problem from a different angle; in trying to find a biological explanation he first attempted to arrive at empirically based conclusions on the relationship between weight and number of nerve fibres and cells on the one hand, and brain weight on the other.[32]

On the basis of measurements of nerve cells produced by G. Levi, I. Hardesty, H. Obersteiner, and H.H. Donaldson among others, Dubois first of all compared the volume of cell bodies in homologous neurons among closely related species of varying size. In the six cases for which data on the measurements of such cells were available Dubois found, comparing the animals pair-wise, that the relationship between body weights P_1 and P_2 and cell-body volumes C_1 and C_2 was $C_1 : C_2 = P_1^r : P_2^r$, in which r was approximately 0.28, in other words half of the exponent 0.56. Dubois then compared the volume of the cell body with that of the neuron as a whole, discovering that the neuron volume increased as the 0.28th power of the cell-body volume. Since the cell bodies themselves increased as the 0.28th power of the body weight, it followed that the neuron volume increased as the 0.56th power of the body weight, that is to say, the same power as for the increase in brain weight in relation to body weight.[33]

Dubois believed that this parallelism justified the supposition that there was a direct link: the increase in brain weight was linked directly to the increase in volume of the sensory-motor units. This also implied that for animals occupying the same level of cephalisation there was no need, moving from smaller to larger animals, to assume a rise in the number of nerve cells. An increase in the volume of the neurons was a sufficient explanation for the differences in brain size.[34]

The second implication was that the nervous system as a whole could now be comprehended as what he called a 'mechanism'. There was simply no room for a quantity of brains, independent of the rest of the body, at the disposal of the 'organ of the mind' or intelligence. Brain size was determined entirely by the quantitative relationships between the constituent parts on the one hand and bodily measurements on the other.[35]

Dubois concluded finally that all the constituent parts of the nervous system had to be seen as functional components. Not only brain cells but also interstitial cells, nutrient cells, and so on, were measured when the weight of the brain was established, and the medullary sheath and the neurolemma were included when neuron volume was calculated. All these accessory tissues contributed to the regularities discovered – they varied in tandem with the nerve fibres and nerve cells, thus constituting a functional whole with them.[36]

Dubois was sure that he had now explained the term $P^{0.56}$, but how was $P^{0.22}$ ($P^{2/9}$) to be interpreted? There were barely any data available on relationships between the sizes of nerve fibres among individuals of the same species. There was some information on the eye but this only led to further complications. According to Lapicque's measurements the retina surface in related species increased as $P^{0.28}$, the exponent being thus half the exponent 0.56. Yet the retina in individuals of the same species did not increase as the 0.11th power – half of the exponent 0.22 – of the body weight, but as the 0.14th power.[37] Dubois left this problem on one side at first, but he quickly produced a drastic solution. In an article in which he delved further into the exponent between human individuals of different sexes, he suggested without further ado:

On the strength of the relations that appear to be valid between the cellular dimensions and volumes in the nervous system, I have become more and more inclined to consider $P^{0.28}$ (more accurately $P^{0.27..}$) as the correct interindividual factor, and to look upon my earlier attempts to interpret $P^{0.22}$ as superfluous.[38]

Dubois did not state his reasons for this step, observing only that the factual material available was insufficient for determining the 'correct' value of the exponent. We can, however, easily reconstruct the circumstances surrounding his decision to work with the exponent 0.28 from thereon, rather than with 0.22, discovered empirically.

Firstly Dubois had established that the cell-body volume also increased as $P^{0.28}$ in individuals of the same species. It was difficult to connect this with the term $P^{0.22}$ for brain increase, but if the increase in brain weight was fixed at $P^{0.28}$, the link immediately became obvious. In individuals of the same species the brain volume would then grow proportionally to the increase in the cell-body volume – thus as $P^{0.28}$ rather than as $P^{0.56}$ as in related species of equal cephalisation. Secondly, the eye factor in individuals of the same species ($P^{0.14}$) was now also half of the inter-individual exponent 0.28, thus strengthening the analogy still further. Thirdly, the exponent 0.28 enabled Dubois to resolve the question of the inter-sexual exponent, as we shall see. This problem was the subject of the article in which he decided to alter the exponent.

Earlier on both Dubois and Lapicque had investigated the exponent between men and women and both had arrived at a value of approximately 0.56.[39] The human sexes thus behaved as related species rather than as individuals from the same species. Neither of the two authors had been able to explain this phenomenon, although it seemed likely that sexual dimorphism played a role. Now Dubois investigated the same problem among various animal species and had to conclude that it occurred only seldom. It did not appear among dogs, nor among sexually dimorphic animals such as cattle and poultry. In all these cases the exponent was closer to 0.28 than to 0.56.[40]

There had to be a cause other than sexual dimorphism, and Dubois believed that he could find it in the differences in muscularity. He showed that men could have up to 27% more muscle weight per unit of body weight than women had, and deduced further that, between men and women, the absolute muscle weight changed in proportion to the square of the increase in body weight. When the exponent was calculated on the basis of absolute muscle weights rather than body weights, the result was 0.28. And if the inter-sexual exponent of 0.56 for the human were truly explained by the difference in the relationship between muscle and body weight, then the value remaining after this difference had been eliminated must be the inter-individual exponent. Here was a new argument for fixing this exponent at 0.28.[41]

In 1922 Dubois was once more to make a radical change in his interpretation of the exponents 0.56 and 0.28, his main reason being a closer study of the relationship between muscle fibres and nerve fibres.[42] The difference in men's and women's muscle weights gave rise to the question of whether the underlying cause was a rise in the number of muscle fibres and the nerve fibres connected to them, or an increase in mass. The latter appeared to be the case. Data on the diameter of muscle fibres indicated that differences in muscle weight among individuals of the same species could be explained entirely by increases in thickness. Dubois concluded that the number of muscle fibres, and thus the number of nerve fibres connected with them, remained the same within the species.[43]

The picture became less simple when one compared individuals from related species. Total muscle mass increased approximately in proportion to body size, but there was no clear link between muscle-fibre thickness and body size. Both large and small animals could possess equally thick muscle fibres and the fibres were even thicker among some small species. Only an increase in the number of fibres could explain this. Yet in that case the number of neurons must also increase, Dubois thought, and thus his original idea that the number of neurons remained the same in closely related species had to be abandoned.[44]

In 1922 Dubois produced a new interpretation, expressed in one breathtaking sentence:

> On the ground... on one side of the relation found for the brain weight E to the body weight P, according to which E increases proportionally to $P^{5/9}$ [$P^{0.56}$] between homoneuric species [i.e., related species of equal cephalisation], and proportionally to $P^{5/18}$ [$P^{0.28}$] between adult individuals of the same species, and on the other side, on the ground of the relation found for the volume of the separate nerve cells C to the body weight, according to which C increases in the ratio of $P^{5/18}$, both between individuals and between species; further on the ground of the established fact that between large and small individuals the number of muscle fibres, hence proportionally that of the nerve cells in the brain, does not differ, but that it differs greatly between large and small species, we may conclude, that also the number of the nerve cells between homoneuric species increases in the ratio of $P^{5/18}$.[45]

Thus Dubois' implicit starting point was the idea that there was a direct link between increase in the volume of the nerve cells and increase in brain weight. This growth in volume was a complete 'explanation' for brain increase among individuals of the same species; there was an additional increase in the number of nerve cells according to $P^{0.28}$ in related animals with an equal degree of cephalisation, the result being an increase in brain weight as $P^{0.28} \times P^{0.28} = P^{0.56}$. Dubois gave this exponent 0.28, which indicated volume increase in both nerve cells and the entire brain within the species, the (somewhat misleading) title 'ontogenetic exponent', because it "expresses the relative individual growth of the brain to the adult state."[46] The value of 0.56, appearing between pairs of different species, was labelled the 'phylogenetic exponent' because it represented "a relation which must have come about with the origination of the species..."[47]

It goes without saying that this model was rooted largely in speculation since Dubois' only empirical justification for his new approach was the fact that, among homoneuric species, the number of muscle fibres must rise along with size increase from species to species. Nothing was known about this rise, in the numerical sense: "the available data do not enable us to *calculate* the exact relation of the body weight to the number of the muscle fibres in these species."[48] What is more, tacitly and with no explanation, he now took as his starting point the idea that the increase in the volume of nerve cells provided a measure of the growth in brain size.

Even though in later years Dubois was to return to 'tinkering' with the ontogenetic exponent, he never made any further revision of his explanatory model.[49] He stopped working on quantitative relations between cell bodies, neurons, and brains, and turned his attention once more to the cephalisation coefficient. The study of this was to lead him to the definitive form of his cephalisation theory.

In the 1922 article introducing the ontogenetic and phylogenetic exponent Dubois posited that differences in cephalisation must be based on divergences, independent of body weight, in the number of brain cells. He believed that a higher degree of organisation could only be attained by one or more extra divisions of the brain cells. He perceived an indication that all, or at least a majority, of the brain cells were involved in these divisions, from the ratio between the cephalisation coefficients of systematically related animals with unequal levels of organisation. Here the ratio of the coefficients usually appeared as simple numbers. For example, for the mouse, hare, and squirrel the cephalisation ratio was 1 : 2 : 3, while the ratio was 1 : 2 for the polecat, stoat, and weasel on the one hand, and stone and pine martens on the other. Ratios of 1 : 1½ : 3 or 1 : 2 : 4 also occurred. To Dubois the only feasible explanation for this regularity was that the brain quantity suddenly increased in an abrupt way by a doubling of all or half the brain volume. The mechanism for this process could only be cell division.[50]

In 1924 Dubois examined the background to this phenomenon.[51] He suggested that whenever brains, or parts of them, had doubled, a "specialisation of the organisation of the animal organic systems" had taken place. He illustrated this with a series of examples, a few of which I quote here:

> In the order of the existing Insectivora the Soricidae [shrews] reach scarcely more than 1/60 of the x [i.e., the cephalisation coefficient] of Man. But a relation, Talpa [the mole], has specialized to get its food by burrowing 'runs' with its large spade-like hands. [He has] reached double the x of that of the families at the lowest stage of the Insectivora... The Bears of the genus Ursus, through their power of climbing and the use they make of their fore-paws as hands, have reached double the value of the x of their lowest existing relations, the Viverridae [civet cats]... The eocene probable ancestor of Elephas, Moeritherium, possessed only 1/6 of the x of its modern descendant, which has developed a very perfect feeler and prehensile apparatus with finger in its proboscis... The Aye-Aye, Daubentonia... reaches double the x of Lemur, through the very remarkable specialisation of its fingers, particularly of the extremely slender and 'ghostly' middle finger, which serves as probe and 'sharp spoon'.[52]

In Dubois' opinion in most cases doubling occurred, which could be explained by assuming an extra division of the brain cells. Since nerve cells no longer divided after birth, this cell division must have taken place in an early stage of embryonic development when the brain was forming. It followed that "development, resting on a single cell-division... takes necessarily place in leaps, 'per saltum', not continuously."[53]

As we have seen, Dubois used only examples in explaining his ideas; there were not enough data available for the formulation of general conclusions on the evolution of the vertebrates, and the data on brain and body weights given in the literature were not always reliable. For example, many measurements were obtained from animals in captivity, and their weight could deviate considerably from that of individuals living in the wild. Also, the precision of the weight measurements often left much to be desired.[54]

It was thus a matter of collecting more, and more reliable data. As we can see from Dubois' correspondence he spent a great deal of time, money, and energy on this in the years that followed. He supplemented his investigation of the literature with his own measurements and with data that were sent to him from all four points of the compass. He usually asked his foreign correspondents to take the measurements for him, according to his own instructions. Whenever possible he also had the animals themselves or casts of their skull cavities sent to him. As well as obtaining measurements of brain weight, he also carried out a great many calculations of skull capacity from which he then deduced the weight of the brain.

In 1928 Dubois judged the time ripe for drawing up general conclusions, at least where the mammals were concerned, and for formulating 'the law of the necessary phylogenetic perfection of the psychoencephalon'.[55] He now finally committed himself to a model of saltatory evolution based upon abrupt brain growth.

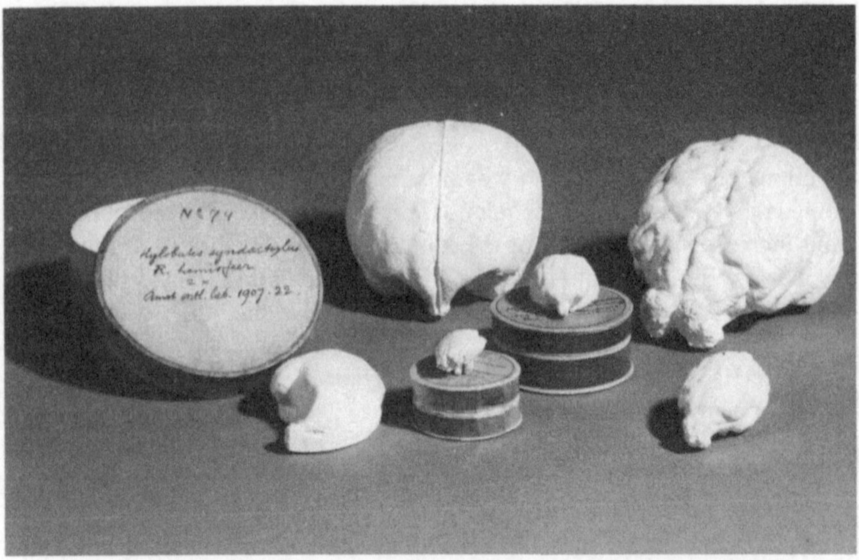

Fig. 16. Endocranial casts, made by Dubois for his cephalisation research.

He began his account with the statement that the results of palaeontology clearly pointed to a discontinuous evolutionary development:

> The absence of proof of continuous development should... also be considered as another important result of the palaeontological research of the earth's crust. The minutest transitions between the members of the many parts of lines of descent with which we became acquainted, were so regularly absent, that this absence can no longer be attributed to 'the imperfection of the geological record'. Every member of a sequence is stepwise distinguished from the preceding and the following one. Again and again we find pillars of the expected bridges, never arches.[56]

He thought he could even go a step further:

> Strictly speaking gradual transition is a priori impossible, because every species living as independent creature, being adapted to particular circumstances of life, must be specialized in its peculiar way.[57]

And elsewhere he added:

> For that matter, the knowledge of the organisms as harmonious wholes, recently obtained through the experimental work of Roux, Driesch, Spemann, and their schools, was incompatible with the earlier existence of incomplete species... [58]

The development of life by abrupt leaps was founded on cell division in the brain, Dubois stated. In his previous calculations he had worked on the total brain weight. However, the brain consisted of functionally very unequal parts; the foremost part, comprising tel-, di- and mesencephalon, could be regarded as "the direct and most highly specialized organ of the psychic functions of the animal."[59] It was this part of the brain, for which Dubois coined the term 'psychoencephalon', that had experienced the most conspicuous increase in volume during the evolution of the vertebrates. If the weight of the psychoencephalon were used in calculating the cephalisation coefficients instead of the weight of the entire brain, then according to Dubois the coefficients fitted into a geometrical progression with ratio 2.[60]

The series for mammals, the class of vertebrates which he investigated in detail, appeared in the following form:

> If the value of the k [i.e., the cephalisation coefficient] of Man is put as unit, the coefficient in the Anthropoid Apes, *Hylobates* included, is 1/4, in the *Canidae*, *Felidae*, and also the *Bovidae* and *Cervidae* differing little from 1/8, in the *Leporidae* 1/16, in the *Muridae* 1/32, and in the *Soricidae*, *Centetidae*, and most *Microchiroptera* 1/64.[61]

Some specialised forms within these large groups could be distinguished by their double or quadruple cephalisation; I have already cited a number of Dubois' examples of this. It appeared that here too, if the calculations were based on the weight of the psychoencephalon, the factor 2 (or one of its multiples) determined the difference in cephalisation.[62]

Dubois was convinced that his model accorded completely with the palaeontological data available. Othniel C. Marsh had shown that brain size among a number of vertebrates had grown very considerably since the Tertiary, and where mammals were concerned he had even talked of 'a general law of brain growth'. He had also drawn attention to the fact that brain growth had mainly occurred in the foremost parts of the brain. In Dubois' opinion Marsh's law was incorrect inasmuch as it posited a brain increase in all mammal groups. Some had remained at a low level of cephalisation, while in other groups there had been several successive duplications.[63]

The law of cephalisation demonstrated that brain volume had not increased as a result of natural selection: "phylogenetic growth..., being autonomous, has a definite direction."[64] Dubois concluded his argument by formulating the implications for the process of evolution:

> Here is a law of evolution [which has] come forth out of the nature of the living being itself, not imposed by the surroundings. Charles Darwin might still think: 'There is no evidence of any law of necessary development' and

Herbert Spencer: 'We find progression to result, not from a special inherent tendency of living bodies, but from a general average effect of their relations to surrounding agencies'. From what we now know of the evolution of the psychoencephalon, i.e. the evolution of animality, it appears that there actually does exist such a law of phylogenesis... with progression, with perfecting. It is self-evident that this perfecting, this steady progression cannot have been caused by factors outside the animal, to which darwinism ascribed phylogenesis. External factors can only have effected diversity of the animal forms and functions, by adaptation. It is inconceivable that they should have acted continually in one definite direction, that of perfection.[65]

In a subsequent article published in 1930 Dubois delved deeper into the way in which new species could have come into being.[66] Phylogeny, like ontogeny, was founded on growth since cell division underlay both processes and the embryo phase was crucial for both. "The principle of phylogenesis" said Dubois "is thus phyloblastesis."[67]

In 1924 he had assumed that cell division in the brain only began after a period of differentiation and specialisation. He made a distinction between vegetative and animal organs, the latter including the nervous and muscular systems; first the vegetative organs, showing a great deal of plasticity, would have differentiated in various directions, whereupon, at a given point in time, the animal system would have developed through an extra division of the brain cells.[68] In 1930 Dubois reversed the order, without explanation, now stating that leaps in cephalisation preceded differentiation in the vegetative organs. The extra division during embryogenesis resulted in a radical change of the animal's organisation. It could be called a macro-mutation, far greater in extent than the mutations that Hugo de Vries had assumed, and at all events incomparable with the gradual changes Darwin and Lamarck had posited.[69]

What happened next, after such a huge mutation? In Dubois' view the mutation in itself only brought about a doubling of the quantity of brains rather than a completely new animal, yet the augmentation created the potential for a total alteration in organisation. To give an idea of how he imagined that this alteration came about, I can only reproduce Dubois' own words:

After birth the mutant selects functions that fit into the changed plan of the form... By 'hologenic somatic induction' of the germ plasma – a developmental process one has to accept and which has been both demonstrated and proved experimentally – the adaptation developed is hereditary and after relatively few generations – certainly an insignificant number when compared with the [future] duration of the new genus' existence –

the adult organism must have attained the characteristic condition which we find in the Earth's past and which we can see before us in the present animal kingdom.[70]

Here the principle of selection operated "usually only in a negative sense, by elimination of the non-viable, sometimes in a positive sense by the mutant's active choice of a functional option."[71]

In 1930 Dubois published an outline article on his cephalisation research in *Biologia Generalis.*[72] In addition to a summary of his results he also provided supplementary information on several points. He argued that the level of cephalisation among many groups of animals did not fit exactly into the series 1 – 1/4 – 1/8 – 1/16... but worked out at figures that were slightly higher or lower. He did not regard these deviations as significant; they found "their apparent cause in minor peculiarities of the great brain in these groups... "[73] Dubois again used examples to illustrate brain growth in leaps, devoting particular attention to animals falling outside the series owing to the fact that, compared with animals closely related to them, they had doubled or redoubled their cephalisation. He also gave a series of examples of cephalisation leaps between extant animal species and their hypothetical Tertiary ancestors.[74]

In this article, as in the previous ones, we search in vain for a numerical foundation to the geometrical series of cephalisation values. There is a total absence of tables of brain and body weights that could be used to calculate the cephalisation coefficients of a particular group. Nor did Dubois investigate, in any numerical detail, the matter of the small deviations he had indicated, providing only examples. Moreover, in a number of cases he only produced ratios between the cephalisation coefficients, without giving the coefficients themselves. Finally, comparison of his publications shows that he sometimes altered the numerical value of the coefficient for a particular group of animals, with no explanation.

To illustrate his conception of the evolution of mammals by abrupt leaps, Dubois used a phylogenetic tree resembling a horse tail (*schachtelhalmähnlicher Stammbaum*; see Fig. 17).[75] We can see that leaps in cephalisation preceded the process of differentiation and specialisation. As proof of this process rather than its inverse (first differentiation, then leaps in cephalisation) Dubois offered the argument – a dubious one since it could also be used to explain the opposite – that species displaying very different specialisations occurred within groups of related animals all with the same cephalisation coefficient.[76]

Once again he maintained that gradual transitions were not possible, and that there was no question of the animal adapting to its environment, but rather of *Einpassung* in the sense used by Von Uexküll:

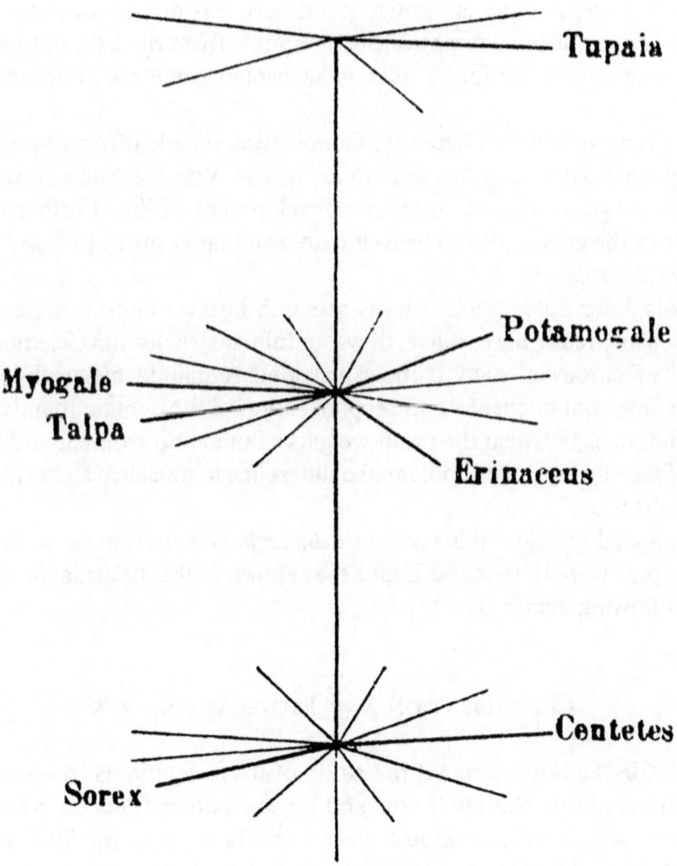

Fig. 17. Example of a *schachtelhalmähnlicher Stammbaum.*

> As one might say a priori, each species, being fitted into [*eingepasst*] its own environment [*Umwelt*] in Von Uexküll's sense, has its own particular specialisation and an organisation perfect in itself; no species is unfinished – such [a species] could not exist.[77]

Adaptation to the environment was achieved autonomously, without outside influences. Each organism accommodated a particular potential which Dubois called the 'growth plan' (*Wachstumplan*) from which the animal could select elements suited to its surroundings. His explanation consisted entirely of examples, one of which is reproduced here:

> For instance, from among the many possible ways of breathing in water, fish have achieved respiration by means of gills, because their growth plan gave the autonomous predisposition towards it. This predisposition is also

present in the mammals' growth plan; here it is not an atavistic heritage, nor evidence of the descent of the mammals from the fish, but only proof that both vertebrate forms have growth-plan elements in common.[78]

There is no doubt that here too, Dubois took his ideas from Von Uexküll, although he did not state this in so many words. Von Uexküll's starting point was also a systematic, autonomous development of life. Furthermore, the example of the gills is also to be found, in a similar context, in Von Uexküll's *Theoretische Biologie*.[79]

Dubois' later publications on his research into cephalisation need not be discussed in detail here, since they contain no really new elements. The amount of empirical data Dubois provided remained niggardly. He only delved somewhat deeper into the cephalisation of the human, mainly to show that differences between the brain weights of men and women, and between those of the various races, indicated a difference in muscle weight rather than in cephalisation.[80]

A promised revision of his article in *Biologia Generalis* never materialised. Other topics were to demand Dubois' attention in the 1930s, as we shall see in the following sections.

Cephalisation and Pithecanthropus

In the 1920s Dubois resumed his study of fossil hominids, beginning with descriptions of the Wadjak fossils and several publications on Neanderthal Man, after which *Pithecanthropus erectus* was brought to the fore again. His renewed interest in Pithecanthropus was partly the result of outside pressures. In 1922 the American palaeontologist Henry Fairfield Osborn had complained, in a letter to the Royal Dutch Academy of Science, about the way in which Dubois had kept his discoveries hidden from the scientific world, rejecting every request to make the fossils available for study. The Academy approached Dubois and elicited an agreement that he would publish new descriptions and illustrations, make casts of all the pieces, and once more allow others to study the fossils.[81]

Dubois' brain research provided another impetus to new study of the fossils. From 1922 onwards, with the renewal of his investigations into cephalisation coefficients, Pithecanthropus once again found a place in Dubois' publications. There was an interaction between the two: Dubois' view of the ape-man's position gave direction to his research on brain growth, while this in turn influenced interpretation of the fossils. Consequently later work on Pithecanthropus must be analysed side by side with the theory of cephalisation.

I shall first discuss several aspects of Dubois' studies of Wadjak Man and the Neanderthals, and then describe the way in which his view of *Pithecanthropus erectus* developed in the 1920s.

We established in the previous chapter that in the first two decades of the twentieth century Dubois associated himself with the phylogenetic approach to fossil hominids. However, after 1920 there was a significant shift of accent in his interpretation, the results of his cephalisation research being the direct cause of this shift. We can see the first indications of Dubois' change of mind in his Wadjak studies.[82]

In his description of Wadjak Man, published in 1921, Dubois ascribed the fossils to a proto-Australian, an ancestral form of the Tasmanians and the Australian Aboriginals. Wadjak Man was distinguishable from these modern races by a receding forehead and a relatively heavy cast to the skull and jaws. His skull capacity was probably larger than that of the Aboriginals. The fossils could not be dated with any certainty, but considering their high degree of fossilisation Dubois regarded the Pleistocene as the most likely period of origin.[83]

Whereas ten years before he had still seen a link between Wadjak Man and Neanderthal Man,[84] he now excluded the possibility of a close relationship between the two forms on the basis of research carried out by the French palaeontologist Marcellin Boule. According to Boule the Neanderthals must be located on a side branch of the human line of descent because morphologically they deviated too sharply from modern Man, while they were also too recent, geologically speaking, to have been his direct ancestor.[85] In Dubois' opinion the Wadjak did have a place on the genealogical line leading to modern humans, and thus could not be regarded as a neanderthaloid form.[86]

In the second part of his account Dubois made a closer examination of the differences between Wadjak Man and the Neanderthals, arguing that the Wadjak was morphologically much closer to the modern human, particularly the Australians, than to the Neanderthals. He explained this by a comparative analysis of the jaws. The Neanderthals' jaw articulations and muscles, he thought, were suited to grinding movements while the jaws of Wadjak Man and the modern human predominantly moved up and down. In his view this indicated a difference in diet. He connected the Neanderthals' jaw structure with the intensive grinding demanded by vegetable food, and concluded that they had been vegetarians. Wadjak Man in contrast must have had an omnivorous eating pattern. These differences in modes of living, with their accompanying adaptations, provided an additional argument for making a sharp distinction between the Neanderthals and *Homo sapiens*.[87]

The most remarkable facet of Dubois' description of the Wadjak fossils is that he deliberately refrained from using expressions such as 'primitive' or 'of low status'. He recorded the distinguishing features of Wadjak Man, modern human, and Neanderthal Man as functional adaptations to their environments, rather than as indications of their level of phylogenetic development. The Wadjak was a predecessor of the Australians but was not therefore automatically more primitive. Rather, the robustness of the fossils indicated to Dubois that Wadjak Man was an 'optimal form' whose living conditions had been more favourable than those of the present-day Australian Aboriginals.[88]

Although there are no direct proofs of this in his treatise, it is likely that Dubois' functional approach was inspired by the steep swing in his ideas on the evolutionary process, resulting from his cephalisation research in these years. In 1922 he made his first open declaration in favour of a saltatory model of evolution. Although it was to be 1930 before he arrived at his final formulation of this model, from 1922 onwards his starting point was that life had evolved in abrupt leaps, in discrete and progressive steps. It seems reasonable to suppose that Dubois' thinking had begun to follow this direction as early as 1920 and 1921, when he was working on the Wadjak fossils.

The explanation of his preference for a functional approach then becomes self-evident. For Dubois the degree of brain organisation indicated the rung reached by an organism on the ladder of evolutionary development. From this it followed that morphological differences between organisms with the same degree of cephalisation could not serve to indicate a different phylogenetic development in a vertical direction, but demonstrated horizontal diversification.

From the description of the Wadjak fossils it is clear that Dubois saw no reason for allotting Wadjak Man a cephalisation coefficient different from that of the modern human; the Wadjak stood on the same level of phylogenetic development and, as we shall see, so did the Neanderthals according to Dubois. In this way a functional approach almost automatically imposed itself on explanation of diversity: differences between organisms with a similar level of cephalisation could best be explained as adaptations to different living conditions.[89]

We see the same pattern in the two studies on the Neanderthals which Dubois published in 1921 and 1922.[90] He set himself the task of demonstrating that the Neanderthals' morphological peculiarities had a largely functional or physiological meaning. In line with Boule's conception Dubois felt that the Neanderthals were too deviant and too specialised to be considered as the ancestors of modern Man, but he disagreed with Boule on the matter of the level of phylogenetic development they had achieved. Boule had described them as archaic creatures, both bodily and mentally inferior to the modern human.[91] Dubois believed, in contrast, that Neanderthal Man's large skull capacity was in itself sufficient proof that he had not occupied a lower position on the human phylogenetic tree. His brain volume was even both absolutely and relatively larger than that of *Homo sapiens*.[92]

Using the results of his cephalisation research, Dubois linked this larger brain volume with a difference in muscle development. Just as men were more heavily muscled than women, so the Neanderthals were more heavily muscled than modern humans, and in both cases this resulted in a larger brain volume. The robust, heavy Neanderthal skeleton corroborated this view.[93] Consequently he argued emphatically against typifying the Neanderthals as primitive or ape-like:

> when we have got to know them better, not a few of these characters will probably be explained as the direct consequence of particular physiological, mechanical adaptations, as phenomena of convergence.[94]

Several features of the Neanderthals' skull provided an example of such mechanical adaptation. In Dubois' view they were directly connected with the position in which the head was held. Boule had argued that the Neanderthals had carried their heads hanging forward and resting on the chest. Seizing on this datum Dubois posited that the skull's form could be explained by the very powerful development of the neck muscles and the epicranial muscular apparatus that must have accompanied this head position. The heavy supra-orbital ridges for instance had served as points of attachment for the foremost epicranial muscles.[95]

The influence of Dubois' cephalisation research is thus even more evident in his publications on the Neanderthals than in that on Wadjak Man. The degree of brain organisation indicated the level of phylogenetic development reached, and in this respect there was no real difference between the Neanderthals and the modern human, so Dubois searched for a functional explanation for their morphological disparities.

From 1922 on *Pithecanthropus erectus* regained its first place in Dubois' interest. In his second study on Neanderthal Man we find the first paragraphs he devoted to the ape-man after many years of releasing almost nothing on the fossils. At the end of this article he pointed out that Pithecanthropus' very gibbon-like skull formation had a partly mechanical background, too, and could not be seen as merely primitive and ape-like. He argued that the Javanese form, like Neanderthal Man, had carried its head hanging forward, and that the skull formation could thus be partially explained by the powerful development of the neck and epicranial muscles.[96]

Shortly after this Dubois brought Pithecanthropus into his research on cephalisation. In the article in which he suggested for the first time that the brain had developed by abrupt leaps, he presented human evolution as an illustration:

> in the Hominides [sic], which geologically do not appear until very late, the brain immediately possessed the same volume already in the earliest

> of the known crania as in modern ones... This however does not apply to Pithecanthropus... But also this 'precursor of Man' is of a late date – probably not before the Pliocene. The transition from such a volume of brain as that of the Anthropoid Apes to the modern human volume seems at all events to have been a rapid one, and halfway there is still that of Pithecanthropus.[97]

Pithecanthropus provided Dubois with one of the motives, if not the main one, for going in search of evidence for evolution by abrupt leaps. As early as 1894, in his first official account of the ape-man, he had drawn attention to the possibility of development 'per saltum'. Now, more than twenty-five years later, he felt able to give concrete shape to such a model.

As shown in the foregoing section, in 1928 Dubois arrived at the conclusion that the cephalisation coefficients of mammals together formed a geometrical series with ratio 2. If human cephalisation was fixed at 1, the coefficient was $^1/_4$ among the apes, $^1/_8$ among carnivores and herbivores, and so on. There was only one hiatus in the series: the factor $^1/_2$ between apes and Man. We can imagine how Dubois filled this gap: he put *Pithecanthropus erectus* there. However, in 1922 he had not yet reached this point. Until 1924 he appeared to avoid discussion of the cephalisation coefficient for Pithecanthropus, restricting his attention to a broad comparison of skull capacities. For instance, he proposed that Pithecanthropus' skull capacity:

> with equal body size (weight), which may be estimated by the femur, [was] double the capacity of the large Anthropoids, and more than two thirds of the cranial capacity of Modern Man (male Australian).[98]

This would indicate a ratio of 3 : 2 : 1, but there seems to be some justification for the suspicion that the coefficients did not bear anything like such a neatly round-figure relation to each other, all the more since Dubois neglected to give the numerical data.

As we can see from the passage just quoted, Pithecanthropus had twice the cephalisation of the apes, but considerably more than half the cephalisation of Man, being closer to the human than to the apes, according to this calculation. This was probably one of the reasons why Dubois decided, when he published new descriptions of the fossils under pressure from the Royal Dutch Academy, to include Pithecanthropus with the Hominidae, a decision he would reconsider later.[99] There is an analogous explanation for another curious passage in this description of the fossils, where Dubois observed that "Man cannot be descended directly from Pithecanthropus, but both [can be descended] from a common primitive Anthropoid."[100] He provided no elucidation of this observation, but it was obviously the uncertainty about the cephalisation ratio between Pithecanthropus and the human being which led him to believe that both forms had derived separately from a line of apes.

Yet this solution obviously did not satisfy Dubois since in 1924 he presented alternative figures, locating the cephalisation of Pithecanthropus exactly half-way between that of the great apes and Man, thus suggesting a direct descent. In this Dubois simply turned his former reasoning on its head. Instead of starting with an estimate of the body weight and using it to calculate cephalisation, he now began with the assumption that Pithecanthropus had half the cephalisation of the human, and argued thereafter that the consequent body weight was entirely acceptable:

> If from the close resemblance of the femur of Pithecanthropus with the human femur the conclusion is erroneously drawn that the organisation of the upper pair of limbs was also the same, x must be put at 2/3 of that of Man, when the body weight is estimated in accordance with human proportions. From the peculiarities of the fossil skull it can, however, be derived that the erect attitude had not yet reached human perfection, and that in consequence of this the upper pair of extremities and with them the upper part of the body, were comparatively heavier than with human proportions. With double the value of x of the Manlike Apes and half that of Man, Pithecanthropus must have weighed about 100 kg. This is a very plausible body weight.[101]

By one stroke of the pen Dubois thus changed the ratio 3 : 2 : 1 between Man, Pithecanthropus, and apes to a ratio of 4 : 2 : 1.[102] The remaining puzzle is how he managed to maintain a ratio of 2 : 1 between Pithecanthropus and the apes or, putting it another way, how he managed to change the ratio between the apes and the human from 1 : 3 to 1 : 4. The matter remained shrouded in Dubois' total silence.

At all events Dubois' further research into cephalisation only strengthened his conviction that he had found the correct solution. In 1924 he established that brain increase usually occurred by doubling, and in 1928 he thought that he could even demonstrate that increases took place exclusively by doubling. Pithecanthropus' cephalisation therefore simply had to be exactly half-way between those of the apes and Man.

I believe that the passage just quoted here provides the key to understanding all of Dubois' later publications on *Pithecanthropus erectus*. In short, the main object of these was to render acceptable the "very plausible body weight" of roughly 100 kg and thus to provide evidence for Pithecanthropus' intermediate position. To this end – although the above quotation seems to suggest otherwise – Dubois directed his particular attention to the characteristics of the femur and to their implications for the ape-man's physical build.

We saw in the previous chapter how Dubois, in his 1900 description of the life-size reconstruction, suggested in passing that Pithecanthropus had not walked upright all the time, but had still possessed a certain capacity for climbing. After 1924 he was to elaborate upon this aspect in particular.

In 1926 Dubois, honouring his promise to the Academy to provide new descriptions and illustrations of the Pithecanthropus fossils, published a treatise on the thigh bone.[103] In this we find clear indications of the direction his ideas on Pithecanthropus' build were taking. In 1922 he had taken the posture of the Australian Aboriginals as a model for that of the ape-man.[104] Now he suggested that "the proportions must have deviated considerably from those of Man, in the direction of those of the Anthropoids, so that the trunk must have been out of proportion longer than in Man."[105] He explained this with the help of a functional analysis of the femur's morphology.

First of all Dubois observed that the shaft of the fossil thigh bone was rather thick in comparison with the human femur, and he believed that greater muscle power and greater body weight could be inferred from this.[106] In his view this larger body weight ought not to be regarded as an accidental, individual feature, but as an indication of a specific physical build and a particular way of moving.

The lower part of the femur shaft was more rounded in section than it was in the human, and this rounding was carried on into the planum popliteum (facies poplitea), which was flatter among humans. There was also no angulus medialis. These features, Dubois stated, had "static and mechanic causes."[107] The adductor magnus muscle had, in this part of the thigh bone, been attached in a different way than in Man, showing that he had functioned differently. The action of the muscle could be compared with that in the apes rather than in humans. The adductor had

> an action going hand in hand with the adducing action, of rotating the femur strongly outwards, or rather an action bringing, with fixed leg, the centre of gravity of the upper part of the body from outside above this leg, and turning at the same time the front to the other side, as in most Apes.[108]

The trochanter major of the femur possessed a feature indicating the same thing, in Dubois' opinion: among humans and most apes it was in a somewhat oblique position with respect to the femur shaft, but in the ape-man it was on an even plane with the shaft. Dubois' inference was that the pelvis must also have been constructed differently from the human pelvis. The explanation for the deviant features in trochanter and pelvis must again be sought in the ability of the thigh bone to rotate outwards.[109]

The implications of the femur's characteristics were, in functional terms, that the ape-man's locomotion "cannot have been exclusively, perhaps not even chiefly, on the ground."[110] Dubois thought that "Pithecanthropus was

Fig. 18. Australian Aboriginal climbing a tree.

less ground-walker than tree-climber, but did not climb with a prehensile foot, in the way of the Apes."[111] He also thought it probable that the ape-man had been able to climb tree trunks just as the Aboriginals did, for example, but without any apparatus. In that case Pithecanthropus would have had very long arms with which to clasp the trunk when climbing; the arms might have served as crutches for movement on the ground. From all this it followed that posture and bodily proportions had been considerably different from that of Man – quod erat demonstrandum.[112]

In 1930 Dubois looked further into the question of what this physical build implied for weight.[113] In his view Pithecanthropus' bodily proportions were more like those of the gibbon than of the human. On the basis of a calculation of the relation between femur dimensions and body weight in humans and apes, Dubois estimated that Pithecanthropus would have weighed 57 kg according to human proportions. Yet taking the gibbon's bodily proportions as a basis the weight would have been 133 kg. The theoretically expected weight of approximately 100 kg, based on a cephalisation coefficient half as large as the human's, thus appeared to be perfectly acceptable, given that the Javanese form was in fact more gibbon-like than human-like in its proportions. In a later article, taking the relation between femur dimensions and body weight of a particularly large male siamang (a species of gibbon) as a standard, Dubois obtained a body weight of nearly 100 kg for Pithecanthropus, which fitted even more closely into his theory.[114]

In 1932 Dubois' assistant at that time, J.J.A. Bernsen, came across a hitherto unnoticed thigh-bone fragment in Dubois' collection of East-Indian vertebrate fossils. Dubois thought this piece could be attributed to *Pithecanthropus erectus*. In the same year further scrutiny of the collection brought three other femora to light which in Dubois' view also belonged to Pithecanthropus. Three years later, in 1935, yet another incomplete thigh bone was to be discovered in the collection.[115] In describing these fragments Dubois argued that they displayed the same characteristic features as the well-known femur, proving that they were species-specific characteristics rather than individual variations.[116] He considered that, with this, "the right of *Pithecanthropus erectus* to exist as a real species, of a separate genus, with a particular organisation, has now been confirmed beyond all doubt."[117]

In 1935 Dubois published a summary of his research into the functional organisation of Pithecanthropus, entitled "On the Gibbon-Like Appearance of Pithecanthropus erectus." It has often been claimed by Dubois' contemporaries and later authors that this article indicated a sudden about-face in Dubois' thinking, and it has been used, quite erroneously, to assert that Dubois had now suddenly begun to see Pithecanthropus as an ape.[118] In reality the article was no more than a repetition, in concise form, of the ideas Dubois had been defending since at least 1924, and partly from 1894 onwards. There was no question of him suddenly describing Pithecanthropus as an ape. On the contrary, he was arguing yet again that where cephalisation was concerned – his only measure for the level of phylogenetic development – the Javanese form stood half-way between apes and human beings and thus at the same distance from both groups.

The misunderstanding must have been provoked by Dubois' portrayal of Pithecanthropus as 'gibbon-like' and 'gibbonoid'. Yet he only meant to show that the ape-man was closer to the gibbons than to Man in build and proportions. In his article he did stress the marked morphological similarity between the skullcaps of Pithecanthropus and the gibbon, but he had drawn attention to this time and again since 1894, and the same obtains for the theory (now once again brought to the fore) that Pithecanthropus must have derived from a line of apes closely related to the modern gibbons.[119]

It can certainly be said that Dubois involuntarily encouraged the misapprehension. In his 1935 article he emphasised the resemblance to the gibbons more than in his 1924 and 1926 accounts of the fossils, while devoting less attention to similarities between Pithecanthropus and Man. This shift in accent arose partly from the simple fact that over the years Dubois had established a number of new similarities between the ape-man and the gibbons, such as the gibbonoid bodily proportions. But apart from this there was another factor, of a polemical kind, playing its part. Dubois' article probably did double duty as a defence document, and we shall see in the next section exactly what it was he had to defend himself against.

Pithecanthropus and Sinanthropus

Dubois believed that Pithecanthropus stood alone on a cephalisation level between the great apes and the human, and was neither anthropoid nor hominid. Whereas he had opted in 1924 for including the fossil in the Hominidae, he later returned to his original idea that only placing the ape-man within a separate family, the Pithecanthropidae, could do justice to the fossil's phylogenetic position.

With this setting there can be little cause for surprise that in the 1930s Dubois should mount a strong resistance to the insights arising from the discovery of *Sinanthropus pekinensis* in China and of new Pithecanthropus fossils on Java. The finds rapidly came to be generally accepted as the final proof that Pithecanthropus belonged with the hominids and was the direct predecessor of all later human forms. For Dubois this conclusion was unacceptable. Either a fossil belonged to the Pithecanthropidae and was thus not a human but an organism with its own typical organisation, or it belonged to the *Homo* forms, which had all achieved the same level of cephalisation as *Homo sapiens* and therefore stood on the same level of phylogenetic development.

The first two skulls of *Sinanthropus pekinensis*, discovered in 1929 and 1930, immediately elicited comments with which Dubois could not possibly concur. The anatomist and anthropologist Davidson Black, who was in charge of the Chinese excavations at Chou Kou Tien and gave an account of the skulls, pointed out the great similarity of form between Sinanthropus and Pithecanthropus. He suggested that Sinanthropus was a more advanced form that could be seen as pre-neanderthaloid, thus occupying a middle position between Pithecanthropus and the Neanderthals. In Black's view there could be no doubt as to the fossils' hominid character.[120]

The German anthropologist Franz Weidenreich concurred with Black that *Sinanthropus pekinensis* narrowed the gap between the Javanese fossils and the Neanderthals, but he thought that the resemblance between the Chinese and Javanese forms was so striking that they had to be considered as one group.[121] The anthropologist Hans Weinert, a compatriot of Weidenreich's, even submitted that the genus name *Sinanthropus* was superfluous, and that one ought to talk of *Pithecanthropus pekinensis*.[122] Marcellin Boule in France adopted the same view.[123]

To Dubois this represented an outright attack on his ideas about the process of evolution and the position of Pithecanthropus, and in 1933 he committed his comments to paper.[124] He agreed entirely with Black and Weidenreich that Sinanthropus belonged with the hominids, but he challenged the idea that Sinanthropus was more primitive than the known human forms, and rejected any suggestion of closer kinship with Pithecanthropus. He conceded that there was a certain degree of morphological similarity between the Chinese skulls and the Javanese specimen, yet comparison of endocranial casts of the skulls showed that the Sinanthropus brain displayed an essentially human feature in the form of a high parietal vertex, which was missing in both Pithecanthropus and the apes. In his view the high vertex was directly linked with the upright carriage of the head: only humans balanced their heads upright above the spinal column, while Pithecanthropus and the apes carried their heads in a slightly overhanging position.[125]

Dubois estimated the capacities of the two Chinese skulls, which Black believed derived from a male and a female, at 918 cc and 1,150 cc respectively.[126] The volume of the male skull was "certainly a very low one for a human skull," he admitted.[127] Yet the individual, who had in all probability been nearly adult, ought to have had a larger skull capacity than the female, in his view, and one should therefore see this as an abnormality rather than as a primitive feature; a male skull of normal development would have attained a capacity of approximately 1,300 cc.[128] His concluding proposition was that *Sinanthropus pekinensis* could safely be regarded as fully human, and ought perhaps to be counted as *Homo neanderthalensis* – "maybe an interesting new race" he suggested magnanimously.[129]

Initially Dubois was not alone in adopting this viewpoint. The American anthropologist Ales Hrdlička also believed that Neanderthals must be involved, while the English anatomist Elliot Smith supported Black's opinion that Sinanthropus and Pithecanthropus were not identical.[130] Yet with the progress of the Chinese excavations and the uncovering of more skulls and other skeletal fragments, the resemblance between the two forms became steadily clearer, and there was increasing support for the view that the fossils belonged to one and the same group of hominids.

In 1937 Franz Weidenreich, who had taken over supervision of the excavations after Black's death in 1934, announced the discovery of a Sinanthropus skull with a capacity of approximately 850 cc, the smallest found up to that time.[131] Other skulls of 1,050, 1,100 and 1,200 cc were found in the same year.[132] Pithecanthropus' capacity thus fell within the range of variation for Sinanthropus, and Weidenreich concluded that "there cannot remain the slightest doubt that Pithecanthropus is a real hominid belonging to the same primitive group as Sinanthropus."[133]

There was no extensive response from Dubois on these later finds, although in 1936 he did revise his previous conception of *Sinanthropus pekinensis* as a neanderthaloid form.[134] Shortly before this, Weidenreich had devoted closer study to the skullcap of the female discovered in 1930 by Black, and had fixed the capacity at 1,025 cc.[135] At this Dubois withdrew his claim that the male skull of 918 cc was abnormal, now arguing that the skull volumes were too small for Neanderthals to be involved.[136]

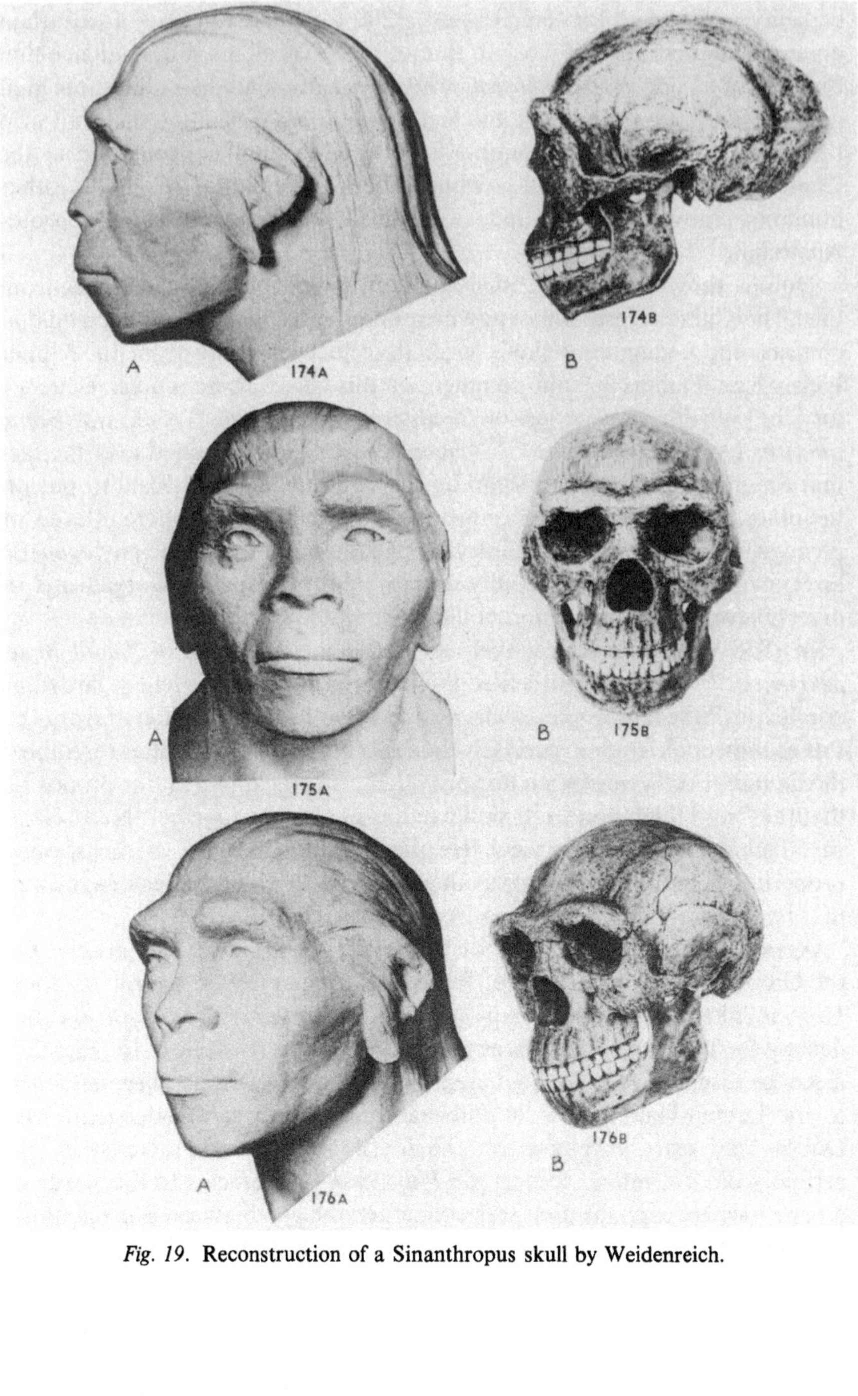

Fig. 19. Reconstruction of a Sinanthropus skull by Weidenreich.

Dubois subsequently declared that the fossils must have belonged to *Homo soloensis*. In 1931 and 1932 eleven skulls from this hominid had been discovered in the Solo valley, close to Ngangdong on Java, by the mining engineer W.F.F. Oppenoorth.[137] According to Dubois this Solo Man was a proto-Australian, comparable to Wadjak Man. He estimated the skull capacity at 1,050 cc for females and 1,200 for males – clearly lower than among Neanderthals but close to that of the Australians. For evidence that Sinanthropus and *Homo soloensis* were the same, Dubois pointed out that not only did both forms have the high parietal brain vaulting, but that also the pars tympanica of the temporal bone had a similar formation in the Chinese and Javanese skulls, while differing from that of all the other hominids known. Sinanthropus was thus, like *Homo soloensis*, a proto-Australian.[138]

Dubois showed very little concern about the problem of age arising from this. The Chinese form was known to be of early-Pleistocene age while in contrast the Ngangdong skulls were described as being from the Upper Pleistocene. Dubois' laconic comment on this was: "There is much evidence for [the] old-Pleistocene age of *Sinanthropus pekinensis*. Would not *Homo soloensis* be of the same age?"[139] He was just as unconcerned over the fact that Sinanthropus and Solo Man on the one side, and Pithecanthropus on the other, were of roughly the same age, despite being on different levels of phylogenetic development. In his view age did not indicate the phylogenetic level; evolution occurred rapidly and by abrupt leaps, and organisms of diverse levels of cephalisation could exist side by side.[140]

In 1938 Weidenreich described several femur fragments of *Sinanthropus pekinensis*.[141] These fossils lacked the features that Dubois had regarded as specific to Pithecanthropus, while on the other hand they differed from the Pithecanthropus femur in precisely those points in which the latter resembled the human. For Weidenreich this provided a strong argument for proposing that the Trinil thigh bone and skullcap did not belong together. Needless to say, Dubois rejected this view. He thought that the lack of the specific properties of the Pithecanthropus thigh bone in the Chinese femora showed that two entirely different forms were involved.[142]

Against this background, Dubois' intention in writing his 1935 article "On the Gibbon-Like Appearance of Pithecanthropus erectus" is now obvious. He wanted to show that the ape-man and *Sinanthropus pekinensis* did not occupy the same level of phylogenetic development. In view of the fact that since the discoveries in China increasingly greater weight had been attached to the human-like features of Pithecanthropus, we can understand why Dubois laid extra emphasis on the fossil's ape-like characteristics. He certainly did not intend to degrade *Pithecanthropus erectus* to the status of an ape – where cephalisation was concerned the gap between Pithecanthropus and the apes was just as wide as that between the fossil form and the modern human; his sole aim was to make it quite clear that there was a sharp division between Sinanthropus and Pithecanthropus.

Meanwhile the first of a series of finds had been made on Java that were to pose an even more serious threat to Dubois' views. Between 1936 and 1940 skull and jaw fragments were discovered at Modjokerto and Sangiran, and the German palaeoanthropologist G.H.R. von Koenigswald, who described them, maintained that they had belonged to Pithecanthropus forms and were at the same time undoubtedly hominid.[143] The situation became even more serious for Dubois when Von Koenigswald and Weidenreich joined forces after initial differences of opinion over the age and interpretation of both their discoveries, and published an article in 1939 in which they posited that:

> Considered from the general point of view of human evolution Pithecanthropus and Sinanthropus, the two representatives of the Prehominid stage, are related to each other in the same way as two different races of present mankind, which may also display certain variations in the degree of their advancement.[144]

Furthermore, in their article both Weidenreich and Von Koenigswald disputed Dubois' thesis that *Pithecanthropus erectus* would have been apelike or gibbonoid. Neither the Javanese nor the Chinese form, they retorted, showed essential similarities to the gibbons; both forms were far above the apes and their proper place was unmistakably within the human family.

Dubois would not hear of it, and in 1936 he began a crusade against Von Koenigswald that was to last until his own death in 1940. Initially his criticism was not entirely misplaced, and until 1939 he was also not alone in his critical attitude, since Weidenreich's verdict at first resembled his own on various points. His recalcitrance resided in his categorical denial, right from the beginning, of any closer kinship between 'his' Pithecanthropus and Von Koenigswald's fossils. Here he was isolated and swimming against the current.

Von Koenigswald's first discovery of an hominid fossil was a child's skull, unearthed in 1936 in the vicinity of Modjokerto, in layers which he believed dated from the early Pleistocene. Initially he attributed the fossil to *Pithecanthropus*, but after protests from Dubois that it was much too human-like to belong to this form – which in any case was not a hominid – he described the find as *Homo modjokertensis* in his official account. Yet it is obvious from the text that Dubois had been unable to convince him, since he described the skull as if it derived from a Pithecanthropus, even though he admitted that there were in fact several differences.[145]

Dubois had absolutely no doubt that Solo Man was involved here. The brain had a high parietal vertex, the position of the ear opening was the same as in the human, and the pars tympanica of the temporal bone displayed a similar structure to that of *Homo soloensis*. As we have already seen, he disposed of the problem of age in an airy manner by suggesting that the Ngangdong skulls could have come from the early Pleistocene.[146]

At first Dubois gained Weidenreich's support for this thesis, the latter feeling that a final verdict could not be given since there could be considerable differences in the features of juvenile and adult skulls.[147] Nevertheless he thought that the fossil resembled Solo Man more than Pithecanthropus; the occiput, for instance, lacked the 'dent' characteristic of Sinanthropus and Pithecanthropus.

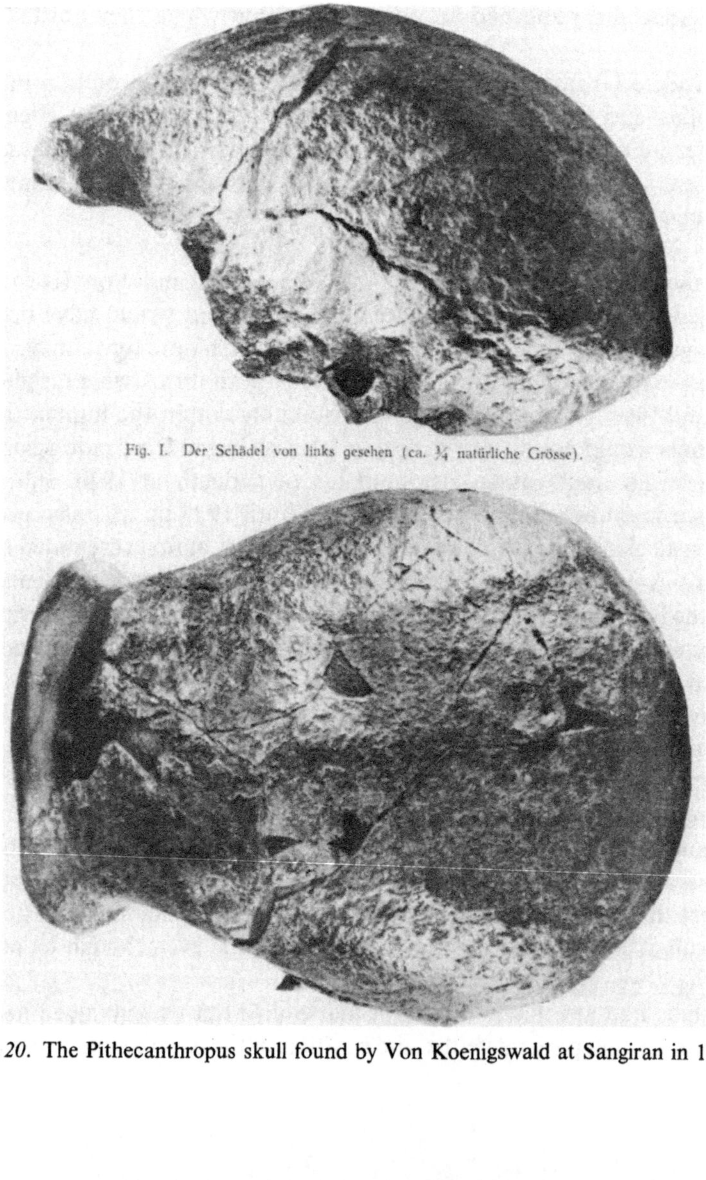

Fig. 20. The Pithecanthropus skull found by Von Koenigswald at Sangiran in 1937.

Dubois was just as merciless towards Von Koenigswald's second find, a jaw fragment uncovered close to Sangiran.[148] Weidenreich also had reservations about it.[149] On the whole the fossil was noticeably more robust than the jaw fragment Dubois had found at Kedoeng Broeboes, and it also displayed several anomalous features. Nevertheless Von Koenigswald attributed it to Pithecanthropus, believing that the differences between this fossil and the Kedoeng Broeboes jaw could be explained by ascribing Dubois' specimen to a female individual, and his own to a male.[150] However, both Dubois and Weidenreich thought that such a heavily built jaw fitted in more easily with the larger Solo-Man skulls than with the relatively graceful Trinil skullcap. Weidenreich did acknowledge that such an interpretation was hardly compatible with the age of the jaw fragment, since according to Von Koenigswald it had come from the same layer as Dubois' ape-man. Later on Von Koenigswald was to claim an even greater age for the specimen and to attribute it to a robustly built Pithecanthropus species that had preceded Dubois' Trinil form.[151]

In 1937 the tide turned in Von Koenigswald's favour when he found a skull, almost complete except for the facial part, at Sangiran. The discovery caused a sensation. Here, a skull belonging unmistakably to the genus *Pithecanthropus* had come to light for the first time since Dubois had carried out his own excavations, and it was moreover in a more complete state than the Trinil find. The back of the head was largely intact, the temporal bones had been preserved, and the positions of the ear and the lower-jaw articulation were easily identifiable. The skull revealed all the typical features to be seen in Dubois' skullcap: the 'dent' at the back of the head, the heavy eyebrow ridges, the postorbital constriction, and the relatively low skull profile. Von Koenigswald's estimate for the skull capacity was 775 cc, even lower than that of the Trinil skull. He believed the fossil to be of the same age as Dubois' specimen.[152]

The special significance of the new find was that it provided convincing proof for the hominid nature of Pithecanthropus. The form of the socket of the lower-jaw articulation and the position of the ear – beneath the zygoma rather than in line with it – left no room for doubt. In these aspects, which distinguished the human from all the apes, Pithecanthropus was completely human.

Yet for Dubois a hominid Pithecanthropus was an impossibility. While acknowledging the morphological resemblance between the Sangiran and Trinil crania, he attached little weight to this, as the Sinanthropus skull discovered first also bore a strong resemblance to the Trinil skullcap. In his view the essential difference was that his 'own' Pithecanthropus was not a hominid, while the hominid nature of Von Koenigswald's find could confidently be deduced from the high parietal vault curve of the brain. His conclusion, as in the two previous cases, was that Solo Man could be involved here, as well.[153]

The discussion acquired a vicious tone when Dubois began to use less elegant means to disqualify the new find. Von Koenigswald had obtained a collection of some thirty fragments rather than the complete skull, yet he had been able to fit the pieces together without much difficulty.[154] Dubois, comparing photographs of the original pieces with the reconstructed skull, now claimed – without taking the photographic perspective into account – that Von Koenigswald had made the skull vault too low. The reconstruction suggested, wrongly, that the specimen was that of a full adult whereas the separate pieces showed that the individual was still immature, he maintained. With this he was accusing Von Koenigswald, indirectly, of incompetence if not actual fraud. He was also making the implicit suggestion that the skull capacity was higher than Von Koenigswald had stated, and moreover that the fossil had not yet reached its maximum growth.[155]

Von Koenigswald was enraged and protested fiercely to the Royal Dutch Academy of Science,[156] whereupon Dubois climbed down a little, declaring that he had not intended to impute dishonest intentions to Von Koenigswald, but he continued to maintain that the reconstruction had been carried out incorrectly.[157] Years later, after Dubois had died, Von Koenigswald would be able to provide definite proof of the accuracy of his reconstruction by means of an X-ray photograph.[158]

Initially Weidenreich was not convinced that a new Pithecanthropus had been discovered either. He drew attention, as Dubois had done, to a striking similarity to Solo Man, and put forward the possibility that the Javanese finds, including those made by Dubois, belonged to female members of *Homo soloensis*.[159] In 1939, however, after Von Koenigswald had been to China to compare his fossils with those of *Sinanthropus pekinensis*, Weidenreich unreservedly sided with his interpretation. As we saw before, both researchers now reached the verdict that Pithecanthropus and Sinanthropus belonged to the same group of hominids. Recognising the strong morphological likeness between Solo Man and Pithecanthropus, they both portrayed the former as a somewhat more advanced later form of *Pithecanthropus erectus*.[160]

Dubois continued to put up a stubborn resistance, the tone of his articles becoming sharper but the force of his arguments waning noticeably. In the last year of his life he collected his grievances into a three-part article, published in the *Proceedings* of the Academy, the last part being handed in two weeks before he died.[161]

In this trilogy he argued once again that all the new finds made in South-East Asia, Sinanthropus as well as Von Koenigswald's Pithecanthropus, and *Homo soloensis* just as much as Wadjak Man, all belonged to the same group of proto-Australians, and had all attained the same level of evolutionary development as *Homo sapiens*. He contended that the differences between the various forms "have only mechanical (direct physiological) significance; they do not correspond to different stages of human evolution."[162] The Trinil fossils were the only exception, and despite all the real or supposed resemblances to the afore-mentioned discoveries, only these fossils derived from that being to which the name *Pithecanthropus erectus* could rightfully be attributed.

Dubois also applied the tried and tested technique to Von Koenigswald's later finds, comprising a fragment from the back of a head, several teeth, and an upper jaw: he attributed them to proto-Australians such as Wadjak Man and Solo Man.[163] The ages Von Koenigswald had allotted to the fossils he dismissed as insufficiently documented since, he said, the fossils were encrusted with material from older layers, making them appear to be older than they actually were.[164]

Dubois ascribed his opponents' conclusions to a bias towards the Darwinian model of evolution. He felt that Darwin's theory had definitely had its day since the theory of cephalisation had demonstrated that evolution occurred by leaps rather than by gradual change.

> Apparently, Von Koenigswald, in his opinion about the true nature of his finds, does rely more on (insufficient) stratigraphical data than on unprejudiced examination of the fossil remains themselves. No wonder that his implicit belief in human evolution from Ape to Man, by gradual transformation of parts of the body, according to the Darwinian point of view, would lead him astray about the true nature of his finds, and wrongly guide the hand which made the restorations.[165]

In the third part of his trilogy Dubois used quotations from his earlier articles to defend himself against statements by Weidenreich and Von Koenigswald that he had begun to regard Pithecanthropus as an ape. He was right on this point; even if he had himself rather encouraged this misconception by ever more emphatically describing Pithecanthropus as 'gibbon-like', Von Koenigswald and Weidenreich had not taken the trouble to absorb the nuances of his viewpoint or the theoretical background to it.

At the end of his argument Dubois once more attacked the model of evolution which his opponents supported. His last words can be held out as an ironical warning to all researchers who are too certain of the objectivity of their scientific work:

> It is most regrettable, that for the interpretation of the important discoveries of human fossils in China and Java, Weidenreich, Von Koenigswald and Weinert were thus guided by preconceived opinions, and consequently did not contribute to, on the contrary they impeded, the advance of knowledge of man's place in nature, what is commonly called human

phylogenetic evolution. Real advance appears to depend on obtaining material data in an unbiased way, such as the *Pithecanthropus* fossils and instructive material about the phylogenetic growth of the brain.[166]

IDEALIST AND PIONEER

In conclusion I shall devote some attention to the historical context and significance of Dubois' work discussed in this chapter. Here I shall be concentrating on the investigation of cephalisation; Dubois' later studies of Pithecanthropus speak for themselves, and they were, moreover, a direct outcome of the brain research.

Dubois' research on cephalisation can best be compared to a reversed pyramid, Dubois having taken an extremely small base of facts on which to build, over the years, a top-heavy superstructure of inferences and conclusions. It is perhaps appropriate to begin with the observation that there is absolutely no reason to doubt either the sincerity of Dubois' intentions or his integrity as a researcher. Everything points to his conviction that his working methods were sound, and there is no trace of deliberate deception on his part. Actually, a look at Dubois' publications is enough to dispel any suspicion of this, precisely because in these he laid bare all his weaknesses. It is also obvious from his correspondence and unpublished notes that he had a profound belief in his work. If Dubois deceived others, he deceived himself just as much.

The research on cephalisation reflected a deep-rooted belief in the order and regularity of the living world. It was axiomatic for Dubois that regular patterns underlay all life's phenomena, patterns that could be formulated with the same mathematical precision as in the laws of physics and chemistry. He expressed discontent with rules-of-thumb or empirical generalisations, seeking continually for biologically meaningful relationships and the underlying biological laws.

Neither the complexity nor the variability of living material could discourage Dubois in his struggle to formulate laws. It was only in his first publication on the cephalisation research that he showed a certain amount of caution over the biological meaning of the relationships he had discovered. Such caution dwindled steadily in the articles that followed, until finally he was ignoring measurements which could not be fitted into his biological explanatory model. For example, the alteration Dubois made in 1918 in the numerical value of the ontogenetic exponent, raising it from 0.22 to 0.28, was based almost entirely on theoretical considerations. Theory became ever more important while the empirical foundation shrunk steadily – the reversed pyramid was beginning to take shape.

The evolutionary model that Dubois developed after 1922 also had an extremely narrow empirical base. An illustration of his belief in order and regularity, in this case the validity of the 'doubling' rule, is the fact that he obviously regarded examples as sufficient proof of his thesis, and did not feel compelled to show reserve as long as a broad empirical basis was lacking.

On the psychological level Dubois' research into cephalisation provides a clear parallel to his journey in search of Pithecanthropus. However shaky the empirical foundation, once Dubois had fallen under the spell of an idea there was no return for him. With a monomaniac obstinacy and the firm conviction that he was in the right, he followed the scent of his conjectures until he had achieved his aim. This very trait, which had contributed considerably to his success in an earlier phase of his career, led him astray in the cephalisation research. In an obituary notice, Arthur Keith hit the nail on the head when he characterised Dubois as "an idealist, his ideas being so firmly held that his mind tended to bend facts rather than alter his ideas to fit them."[167]

There was never much support for the evolutionary model Dubois developed from his research into cephalisation The Amsterdam physician Reindert Brummelkamp was one of the few researchers to follow in his footsteps and achieve similar results. Like Dubois, Brummelkamp found a value of $5/9$ for r in the formula $E = cP^r$. He also thought he could establish that the coefficients of cephalisation in the mammals together formed a regular series, but rather than an increase in the coefficient by a factor of 2, he found leaps by a factor of $\sqrt{2}$.[168] Dubois' ideas gained support from the zoologist J. Versluys, while the anatomist H.J. Bijlmer and the histologists A.B. Droogleever Fortuyn, S.T. Bok, and M.J. van Erp Taalman Kip adopted a sympathetic attitude to his model.[169] The anatomist C.U. Ariëns Kappers used the factual part of Dubois' work in his own research into the brain, but remained carefully noncommittal towards Dubois' evolutionist inferences.[170] As far as I have been able to ascertain, Louis Lapicque was the only foreigner to side with Dubois,[171] otherwise there were only dismissive reactions from abroad.

Criticism of Dubois' ideas erupted in the second half of the 1930s, with unfavourable comments not only on his model of saltatory evolution but on the cephalisation research as a whole. To a large extent the criticism was justified. In 1937 Gerhardt von Bonin demonstrated that the whole undertaking had been doomed to fail right from the beginning because a completely arbitrary premiss had been taken as the basis for calculating the phylogenetic exponent. Dubois had discovered a value of $5/9$ for the exponent by making a comparison between pairs of animals assumed to have an equal level of cephalisation. Von Bonin rightly pointed out that this procedure was unacceptable. It had to be shown that particular animals did actually have the same degree of cephalisation, and the mere assumption that they did could not be used as a starting point. Von Bonin established that the phylogenetic

Fig. 21. Dubois at the end of the 1930s.

exponent acquired a value of approximately ²/₃ when the calculation was based on all the data available rather than on pairs selected in an arbitrary way. Calculation of cephalisation coefficients using this exponent showed that there was no question of a geometric progression with a ratio of 2.[172]

In 1939 Raoul Anthony also criticised Dubois' procedure, although for different reasons. In his view body weight was too variable a factor to produce reliable results. He showed that the phylogenetic exponent and the cephalisation coefficient varied widely, depending on the body weight assumed for an animal species. Anthony believed that this raised doubts not only about the accuracy of the value 5/9, but also about Dubois' thesis that there was actually a universal exponent for the mammals. He also felt that the structure of the brain was too complex for an unequivocal significance to be attached to the cephalisation factor.[173]

In 1948 David Sholl revealed another weak point in Dubois' and Brummelkamp's approach: the complete lack of any form of statistical testing of the results. Sholl believed that nothing more than conjecture underlay Brummelkamp's thesis that the cephalisation coefficient increased by 'doubling' leaps: "in fact, Brummelkamp merely says that if you plot the lines he suggests, the data group themselves about these lines."[174] There was of course even more reason for applying this criticism to Dubois' model. Sholl also demonstrated that, from a statistical viewpoint, the data upon which Dubois and Brummelkamp had based their calculations simply did not justify their conclusions.

In the 1940s comments similar to those made by Sholl, Anthony, and Von Bonin were published by E.W. Count and K. Wirz, among others.[175] Most of the criticism appeared after Dubois' death, and he did not respond to the articles by Von Bonin and Anthony in 1937 and 1939. Dubois lived in seclusion on his Haelen estate in the last years of his life, and it is possible that he never even saw these publications. On the other hand he was certainly familiar with Brummelkamp's work, but Brummelkamp was an ally rather than an opponent. Even though he found a different multiplication factor, his research left intact the most fundamental premisses of the cephalisation theory. Apart from that, Dubois refused to give recognition to Brummelkamp's conclusions, believing that the other's anomalous results were entirely due to a lack of scholarly judgement in the collection of data on weight.[176]

In contrast to what one might expect, given the crushing criticisms, Dubois' efforts were not entirely fruitless. What can be said of his activities within the field of palaeoanthropology also applies in essence to his research into cephalisation: it was pioneering work. Dubois' strength lay in the originality of his approach, in his ability to open up new paths of research. In his studies of the relationship between brain and body weights he was one of the first to draw attention to the phenomenon of allometric growth and to try to analyse it by quantitative and mathematical means.[177] Otto Snell had provided the first impulse for this with his formula; Dubois recognised its significance for a new approach to morphological problems and produced the first calculations.

Louis Lapicque and his supporters followed in Dubois' footsteps. In 1919 Bernhard Klatt drew attention to the potential for new applications, giving the example of the relationship between body weight and the weight of the heart.[178] In 1925 Raoul Anthony, who later turned into a critic of cephalisation theory, tried to apply Dubois' formulae to brain growth during ontogeny.[179] In the 1920s and 1930s Julian S. Huxley and G. Teissier employed an allometric perspective to examine, among other things, embryological growth. In 1932 Huxley gave a survey of allometric phenomena in his *Problems of Relative Growth*. Particularly after the publication of this book, in which he also mentioned the studies by Dubois and Lapicque, the allometric approach rapidly gained acceptance.[180]

Although the criticism of Dubois' work was largely justified, it also had a less favourable effect in its 'overkill'. For some time after the excesses in Dubois' and Brummelkamp's work came to light, many scientists shied away from allometric studies of brain and body weights; they had turned into dangerous ground. In the 1950s and 1960s however, Harry J. Jerison in particular mounted a spirited defense of the research. Since then the resistance has gradually worn off, and studies in this area have once again increased in number.[181] Needless to say, everyone undertaking such a study goes about it in a far more cautious and empirical way than Dubois, since the field continues to be full of pitfalls and traps. There is still great uncertainty about the biological meaning of the exponents; the numerical value of what Dubois called the phylogenetic exponent is still disputed; there is the view that Dubois' cephalisation coefficient is not independent of body weight; and the idea of the multiplicative development of the brain has been abandoned and replaced by the assumption that enlargement of the brain is based on addition of new parts.[182]

Allen, in his *Life Science in the Twentieth Century*, states that:

> Biologists in the 1880s or 1920s continually looked to physicists and chemists for models of how scientific investigations should be carried out. Biologists were aware of the truth of the accusation, often levelled at them by workers in the physical sciences, that their field was not scientific – that they could not rigorously prove their conclusions and that many of their lines of evidence were incomplete and tenuous.[183]

Dubois' struggle to express his ideas in mathematical and quantitative terms, no matter how primitive, and his conviction that living phenomena were governed by laws having the same status as those of physics or chemistry, easily fits into the tendency Allen has indicated.

Allen also observed that the first generation of biologists to adopt the methods of physics and chemistry as their ideal often had too little grounding in these fields to be able to put their ideas into practice with any success.[184] According to Allen it was only in the 1930s and 1940s that researchers with a thorough training in the natural sciences moved into the field of biological investigation. The parallel with Dubois' work is obvious here, too. He tried to achieve mathematical precision but his background knowledge fell sadly short. It was to be a younger generation of biologists who, armed with Ronald A. Fisher's works for example, would make short work of Dubois' 'exact' approach.

With the benefit of hindsight we can see that the whole structure of Dubois' cephalisation research had to be dismantled to ground level before anyone could begin to build anew, yet at least part of the foundations laid by Dubois remained intact. At all events with his attempt at quantification he had introduced a new element into morphological research.[185]

A second innovative element discernible in Dubois' work is his plea for functional morphological research. He realised that morphologists had concentrated too much on analysis of form, neglecting the functional aspect. In his own cephalisation research he frequently made use of physiological insights into the construction and function of the nervous system. He also tried to explain his results from a functional angle, connecting differences in the degree of cephalisation with differences in the specialisation of animal functions. His model of evolution by leaps gave him an even clearer insight into the importance of the functional perspective, as one can see from his studies of Neanderthal Man and his analysis of the Pithecanthropus femur, for example.

Several of Dubois' statements show that he consciously distanced himself from the exclusive preoccupation with aspects of form that was typical of the German school of morphology. For instance, in one of his later articles on Pithecanthropus he observed that "the physiological method of research thus taught [us] to understand the principal properties of this organism, where the purely morphological method was almost powerless."[186] In a letter to the anatomist J.H.F. Kohlbrugge he even suggested that Haeckel had come to his monism via one-sided attention to aspects of form:

> The exclusive consideration of form... to which Haeckel and his supporters give precedence, has resulted in material monism. Only a many-sided view, and a physiological one at that, can free us from this monism. Only by this means can real laws be traced in living nature as well... [187]

There was a direct link between Dubois' interest in functional aspects and his research on cephalisation, but he nevertheless was not alone in his efforts to rehabilitate the functional approach, long neglected, within morphology. Starck has suggested that examples of a renewed interest in functional aspects can be found in the work of Fürbringer, Dubois' teacher, and also in that of the zoologist Max Weber. In the 1920s and 1930s K. Peter, A. Benninghoff, and H. Böker, among others, propagated the functional approach in Germany,[188] and C.J. van der Klaauw and his school did so in the Netherlands.[189]

In the British case Zacharias has shown that the anatomists Arthur Keith, George Elliot Smith, and Frederick Wood Jones tried to reconstruct the phylogeny of the primates, including humans, by means of functional analyses.[190] For instance, Keith explained the human's upright gait by a series of changes in the use of the lower limbs, which in his view must have occurred in the course of the primates' evolution. Using this succession of functional stages as his basis, he determined which groups of primates could be fitted into the human family tree, and from which line of apes the human must have derived.[191] While the classical morphologists took form as their basis for constructing a genealogical tree, these British anatomists took function as their starting point. The fact that their phylogenetic trees were often just as speculative as those drawn up by Haeckel and his followers – Zacharias talks of speculative functional phylogenies – is of less importance here; what matters is that they began their investigations with a different perspective.

Even though these examples of the functional approach within morphology vary considerably as to the questions raised, the common factor between them is that they are clearly distinguishable from the German school with its exclusive concentration on analysis of form. Here new paths were being hacked out for morphological research, and in my view this provides an interesting area for historical study, especially since historians of science have paid little attention to the post-1900 development of morphology up to now, certainly when compared with the interest for the experimental disciplines.[192]

As we observed in an earlier chapter, in his publications Dubois rarely commented on questions of scientific theory and philosophy, believing that, as a natural scientist, he ought to keep well away from these areas. For example he never formulated, in so many words, the belief in the orderly pattern of the natural world which speaks so clearly through his work. Because of his reticence it is impossible to provide a detailed picture of the ideas behind his work, and it is precisely his aloofness that gives one reason for assuming an absence of any carefully-thought-out viewpoint on his part. I would nevertheless like to end with an attempt to map out some of Dubois' theoretical and 'philosophical' notions.

In his studies of the brain, particularly those of 1918–1920, Dubois tried to achieve a fully mechanistic explanation for the ratios between brain and body weights, and between the constituent parts of the nervous system. He believed that the brain could be seen as a 'mechanism'. There was no need to take immaterial factors or mental aspects into account. Dubois never went so far as to adopt a purely materialist standpoint; he never explicitly denied the existence of the non-material. For him this was a matter on which the natural scientist should not comment, and he maintained an agnostic attitude towards questions which he deemed to be beyond the reach of the methods of scientific research. This can be illustrated by the following quotation from a letter to Kohlbrugge. Referring to one of his articles on his brain research, Dubois observed:

> The [article] ends by indicating the wholly mechanical construction of the nervous system. By that is also meant "that all psychological phenomena are directly determined by morphological structure, [that they are] a result of it," which is an axiom you object to in your "Morphologische Abstammung des Menschen"... I have come round to thinking this on the strength of proper facts and I believe that it was not wrong to take this axiom as a starting point. Yet I do *not* mean that "all dualism is excluded beforehand"... To trace all material occurrences back to phenomena that are subject to general mechanical laws, in the nervous system and the brain as well, no more conflicts with the dualistic view than do Kepler's laws with the existence of a Supreme Being.[193]

In the 1920s Dubois' research into cephalisation steadily took him further away from Darwinism. After 1920 he also slowly but surely distanced himself from his previous mechanistic position. For instance, in one of his articles he called the autonomous growth of the brain "a fact that lends firm support to the idea that evolution within the animal kingdom is just as inexplicable in its origins as is life itself."[194] He also observed in a letter: "Happily I have lived to see... the mechanical ideology making way for a more accurate [one], especially where Man is concerned."[195]

However, the content of this "more accurate" view remains uncertain. The fact that in his later articles on the cephalisation problem Dubois introduced such concepts as 'Einpassung' and 'Wachstumplan', borrowed from Von Uexküll, may give rise to the idea that he associated himself with idealist biology. His starting point in these articles was indeed an autonomous development of life according to a metaphysical plan that was only partially amenable to causal analysis. Yet there is no fuller elaboration of these ideas to be found anywhere in Dubois' works, and after 1930 he did not even return to Von Uexküll's ideas at all.[196]

In the last ten years of his life Dubois took a strong interest in the holistic movement arising within biology in the 1920s and 1930s. This trend originated to a large extent in experimental disciplines such as physiology and experimental embryology. Quite a number of researchers became convinced in this period that a reductionist approach was inadequate to explain such phenomena as induction, differentiation, regeneration, and homeostasis. It was argued that the development and functioning of the organism could not be fully comprehended by analysis of its separate parts, and that it was necessary to give attention to the organism as an integrated whole. The holists believed that the whole possessed features which could not be explained in terms of the structure and action of the separate parts. Among the well known supporters of such a 'Ganzheits' conception of the organism were Hans Driesch, Walter B. Cannon, Lawrence J. Henderson, Joseph H. Woodger, and E.S. Russell.[197] In the Netherlands holistic ideas were advanced by the neurohistologist Jan Boeke, the physiologist Hermann J. Jordan, and the morphologist C.J. van der Klaauw, among others.[198]

For Dubois, holism had anti-Darwinist implications, since the fact that an organism formed an harmonious and appropriately functioning whole proved in his view that evolution could not possibly have been a gradual process. Every arbitrary interference or change in the organisation of the organism would disturb the harmony of the whole, thereby destroying the viability of the organism. Only a radical and purposeful reorganisation could transform a functioning whole into a new 'Ganzheit'. The accidental and arbitrary variations assumed by Darwinism would produce an organism that was "unfinished", and "such [an organism] could not exist."[199]

Of course, this was not a new criticism, since Hans Driesch for example had advanced the same arguments against Darwinism as early as 1894.[200] Besides, ever since the first appearance of the *Origin of Species* critics of evolutionary theory had, without recourse to holistic concepts, time and again raised the objection that blind chance could not bring forth viable new organisms.

Dubois believed that the results of his cephalisation research fitted perfectly into the new holistic outlook. For example, in a letter to G.A.F. Molengraaff, dated 11 October 1930, he stated:

> What we have learned from the phylogenesis of the central organ of the animal functions is just as incompatible with the current assumptions about animal organisation, developed largely under the influence of Darwinist notions, as what we learned from the pioneering investigations of the German experimental schools. I do not need to tell you that I am gratified by this... because the cephalisation research thus helps to bring about the new biological phase.

Weidenreich and Von Koenigswald, however, had completely missed the boat:

> The insufficiency of [their] endeavours... to prove in China and Java the reality of human evolution in [the] Darwinian way, by gradual, slow and steady transformation of parts of the body, is a consequence of their neglecting the holismic or *Ganzheits* conception of the organism, firmly established by the researches in experimental embryology, on which conception, although having their origin early and spontaneously in an-

other starting-point, are also based on my own and L. Lapicque's researches on phylogenetic progress of the brain, especially the cerebrum, in the Vertebrates.[201]

Nevertheless, Dubois did not develop a detailed holistic view. Just as he used a number of Von Uexküll's concepts without thereafter turning into an idealist biologist, so he took only general notions from holism without formulating a clearly defined holistic standpoint. In this he reveals himself as an eclectic, taking from current ideas and theories whatever fitted the results of his research, without bothering to go deeper into their theoretical background.

Notes

1 In recent literature the term encephalisation is used instead of cephalisation. I shall stick to the older term here.
2 See chapter 3, p. 63 and chapter 4, p. 93.
3 Weber, "Vorstudien über das Hirngewicht"; "Over het hersengewicht".
4 Dubois, "De verhouding van het gewicht", 1897.
5 *Ibid.*, 1–10.
6 *Ibid.*, 27–28. Dubois based this on Snell, "Das Gewicht des Gehirnes"; "Die Abhängigkeit des Hirngewichtes".
7 Dubois, "De verhouding van het gewicht", 1897, 10.
8 *Ibid.*, 15.
9 *Ibid.*, 14–26.
10 *Ibid.*, 10–14.
11 *Ibid.*, 21–26.
12 *Ibid.*, 26–27.
13 *Ibid.*, 29.
14 *Ibid.*, 30.
15 *Ibid.*, 35.
16 *Ibid.*, 34–35.
17 Dubois, "The Brain-Cast of Pithecanthropus", 1899.
18 *Ibid.*, 85–86.
19 *Ibid.*, 86–96.
20 *Ibid.*, 96.
21 In 1898 Dubois also investigated the relationship between brain and body weight in humans ("Ueber die Abhängigkeit des Hirngewichtes", 1898). The stimulus for this had been a series of measurements of men sent to Dubois, after the publication of his first article, by the German anthropologist Otto Ammon. The results of this research, which Dubois later resumed and extended, will be discussed further on.
22 Dubois, "On the Relation between the Quantity", 1914.
23 Lapicque and Girard, "Poids de l'encéphale"; Lapicque, "Le poids de l'encéphale".
24 Dubois, "On the Relation between the Quantity", 1914, 655.
25 Lapicque, "Le poids encéphalique"; "Sur la relation"; Lapicque and Dhéré, "Sur le rapport".
26 Dubois, "On the Relation between the Quantity", 1914, 655–658. In his first article, in 1898, on the exponent between individuals of the same species ("Ueber die Abhängigkeit

des Hirngewichtes", 1898) Dubois had only included humans in his investigation (compare with note 21). At that time he attributed the low value of the exponent to the fact that the human brain was strongly convoluted. Lapicque had likewise considered the low value in dogs to be an isolated case (Lapicque and Dhéré, "Sur le rapport", 765). Now that the exponent was seen to be a general one, these explanations lost their validity in Dubois' view.

27 Lapicque, ("Tableau général", 260–261) expressed it in the following way: "Dubois' law [which is] a general, empirical law, seems to us to be the limit of the sum of several different functions; it is only the exact knowledge of each of these functions, studied separately, which will enable us to give a rational solution to the problem."

28 Lapicque, "La grandeur relative"; "Relation du poids".

29 Dubois, "On the Relation between the Quantity", 1914, 660–666.

30 For example, in the English version (*ibid.*, 661), as well as in the Dutch one ("De betrekking tusschen hersenmassa", 1914, 607) Dubois suggested, in explanation of the factor L, that "if we admit that to every nerve-fibre a definite central cell-mass answers, then these masses must increase with the number of nerve-fibres, in segmentically constituted [i.e., metameric] animals indeed as the longitudinal dimension. Yet in the German version ("Die gesetzmässige Beziehung", 1914, 342) we read: "If a certain amount of excitation energy corresponds to a certain central cell mass, then this mass must grow according to the diminishing working efficiency, thus in metameric animals according to the length dimension of the body." The quantity of excitation energy was in Dubois' view determined by the mass of the nerve fibres; he thus correlated the increase in the size of the brain with increase in the mass of the nerve fibres, and not with increase in their number. However, when explaining the eye factor, in his German article as well, his starting point was a correlation with the increase in the number of nerve fibres.

31 Dubois, "On the Relation between the Quantity", 1914, 666–667.

32 Dubois, "On the Relation between the Quantities", 1918; "The Significance of the Size", 1919; "Comparison of the Brain Weight", 1919; "The Quantitative Relations", 1920.

33 Dubois, "On the Relation between the Quantities", 1918, 1332–1336.

34 Dubois, "The Significance of the Size", 1919, 725. Dubois did not immediately draw this conclusion. Initially he did actually use an increase in number as his starting point ("On the Relation between the Quantities", 1918, 1336).

35 *Ibid.*, 726. See also Dubois, "The Quantitative Relations", 1920, 665–667. In 1885 Manouvrier had defended the existence of an independent brain quantity that was linked with intelligence ("Sur l'interprétation de la quantité", 316). Lapicque ("Tableau général", 255–257) had challenged this thesis in 1907 by pointing out that the formula suggested by Manouvrier for determining this independent brain quantity resulted in incongruities when applied to concrete cases. Dubois sided with this criticism ("The Quantitative Relations", 1920, 666–668.). Further Dubois also thought he could demonstrate that the relations between the constituent parts of the nerve cells were quantitatively determined (*ibid.*, 671–675.) He was able to calculate, from measurements of cells and their contents, that the plasma volume increased in proportion to the length dimension of the body, i.e. as $P^{1/3}$ or L. The volume of the nucleus varied as $P^{1/6}$. Thus there was also a fixed relation between nucleus and plasma. Dubois concluded that: "The existence of these fixed relations of volume of the neurone and its parts leave, indeed, hardly any room for doubt of the perfectly mechanical character of its arrangement" (*ibid.*, 676–677). In his view the relations discovered were in complete agreement with what was known about the physiological significance of the plasma and the nucleus. For further details, see *ibid.*, 677–680; "Phylogenetic and Ontogenetic Increase", 1923, 253–255.

36 Dubois, "The Quantitative Relations", 1920, 665–666.

37 Dubois, "The Significance of the Size", 1919, 711–715. Dubois suggested in this article (*ibid.*, 712) that the eye factor was directly linked with the increase of the cross-section

of the nerve fibres, but in later publications he relinquished this idea without comment.

38 Dubois, "Comparison of the Brain Weight", 1919, 850.

39 Lapicque, "Le poids encéphalique"; "Comparaison du poids encéphalique". Dubois, "On the Relation between the Quantity", 1914, 658–659.

40 Dubois, "Comparison of the Brain Weight", 1919, 850–855.

41 *Ibid.*, 858–869. As a cross check Dubois also calculated the relation between body and muscle weights in various animal species for which an intersexual exponent of 0.28 was valid. Here, as expected, the increase in muscle weight was proportional to the increase in body weight (*idem*).

42 Dubois, "Phylogenetische en ontogenetische toeneming", 1922; the English version ("Phylogenetic and Ontogenetic Increase") was published in 1923.

43 Dubois, "Phylogenetic and Ontogenetic Increase", 1923, 244–245.

44 *Ibid.*, 245–247.

45 *Ibid.*, 247.

46 *Ibid.*, 241.

47 *Ibid.*, 234.

48 *Ibid.*, 247.

49 In 1930, when he published a survey of his cephalisation research ("Die phylogenetische Grosshirnzunahme"), Dubois was still working with the same exponents as in 1922. In 1935 ("Pithecanthropus als organisme", 151) however, he suggested in connection with the ontogenetic exponent that: "By comparing eleven available organisms 0.27 was found as the average value of r, thus a little more than 2/9. It can be assumed that an average r = 2/9 will also be found for a great many organisms." In 1938 ("Over de spronggrootte", 38) he wrote: "Between individuals of one and the same species the ontogenetic exponent is approximately 2/9..." Thus Dubois obviously abandoned the exponent 0.28 in later years, and returned to the value of 0.22. His motives for this step are unclear, and neither did he discuss its consequences for the model drawn up in 1922, based on the exponent 0.28. Apart from this, the lack of certainty about the proper value for the ontogenetic exponent had no direct consequences for the model of saltatory evolution that Dubois developed on the basis of his theory of cephalisation.

50 Dubois, "Phylogenetic and Ontogenetic Increase", 1923, 247–249.

51 Dubois, "On the Brain Quantity", 1924.

52 *Ibid.*, 432–434.

53 *Ibid.*, 432, 435–437. The quotation is taken from p. 436.

54 *Ibid.*, 431–432.

55 Dubois, "The Law", 1928.

56 *Ibid.*, 304.

57 *Idem.*

58 Dubois, "Phyloblastese", 1930, 126.

59 Dubois, "The Law", 1928, 305.

60 *Idem.*

61 *Ibid.*, 309.

62 *Ibid.*, 310–311.

63 *Ibid.*, 306–308.

64 *Ibid.*, 313.

65 *Ibid.*, 313–314.

66 Dubois, "Phyloblastese", 1930.

67 *Ibid.*, 124.

68 Dubois, "On the Brain Quantity", 1924, 437.

69 Dubois, "Phyloblastese", 1930, 125–126.

70 *Ibid.*, 126–127.

[71] *Ibid.*, 127.
[72] Dubois, "Die phylogenetische Grosshirnzunahme", 1930.
[73] *Ibid.*, 259.
[74] *Ibid.*, 259–274.
[75] *Ibid.*, 284.
[76] *Ibid.*, 284–285.
[77] *Ibid.*, 286. For further information on Von Uexküll's concepts *Einpassung* and *Umwelt*, see for example his *Umwelt und Innenwelt der Tiere* and his *Theoretische Biologie*, chapter 8, especially 226–233. See also Schmidt, *Die Umweltlehre Jacob von Uexkülls*, chapter 2.
[78] Dubois, "Die phylogenetische Grosshirnzunahme", 1930, 290.
[79] See Von Uexküll, *Theoretische Biologie*, chapters 7 and 8, especially 188–195. The example of the gills can be found on p. 193–194. See also Schmidt, *Die Umweltlehre Jacob von Uexkülls*, chapter 4.
[80] Dubois, "The Seeming and Real Cephalisation", 1933; "Ueber die Gleichheit", 1934. Dubois fixed the human ontogenetic exponent at 0.24 instead of 0.28. In his view the exponent for domesticated species was generally lower than the exponent for wild animals. Compare also Dubois, "Das Gehirn beim Haushunde", 1922. Although Dubois published nothing more on the subject after 1930, in later years he continued to concern himself with the problem of how new species came into being. It is clear from unpublished notes that he was studying Lamarck's works, and Lamarckism, around 1935. In a letter of 1934 to the German palaeontologist Tilly Edinger he even observed without further elucidation that the investigation into cephalisation "[speaks] unambiguously in favour of Lamarck." He may have meant here that the process of *Einpassung* could take place according to Lamarckian principles. However, it should not be supposed that Dubois became a Lamarckist since there are no references to Lamarckism in his notes or publications from later years. Further it can be gleaned from unpublished notes that Dubois began, in the 1930s, to have new doubts about the order in which mutation and differentiation took place, and was inclined to return to his original idea of 1924. The reason for this is unclear.
[81] Copies of the correspondence connected with this question can be found in the Dubois Collection.
[82] Dubois, "The Proto-Australian Fossil Man", 1921. Keith suggested, probably correctly, that Dubois was stimulated to produce his description of Wadjak Man by the account, published by S.A. Smith in 1918, of a fossil human skull from Talgai in Australia. (*Antiquity of Man*, 440).
[83] Dubois, "The Proto-Australian Fossil Man", 1921, *passim.*
[84] Compare with chapter 4, p. 119–120.
[85] Boule explained his ideas on Neanderthal Man in his "l'Homme fossile de La Chapelle-aux-Saints". See also his *Les hommes fossiles*, chapter 7.
[86] Dubois, "The Proto-Australian Fossil Man", 1921, 1018–1020.
[87] *Ibid.*, *passim.*
[88] *Ibid.*, 1030.
[89] Perhaps superfluously: of course, Dubois did not exclude the possibility of primitive characteristics occurring, but they were less important to him than to the supporters of a gradualist model of evolution, on the one hand because, for him, only the degree of cephalisation was indicative of 'vertical' phylogenetic development, and on the other hand because the diversity within a group of animals of equal cephalisation was determined by the non-primitive features.
[90] Dubois, "On the Significance of the Large Cranial Capacity", 1921; "On the Cranial Form", 1922. In the latter article Dubois also described the fossil hominid skull found in 1921 at Broken Hill in Rhodesia. He considered this, like Wadjak Man, to be a proto-Australian form (*ibid.*, 325–331).

[91] In his *Les hommes fossiles* (242) Boule described the Neanderthals in the following way: "That *Homo Neanderthalensis* is a species with archaic characteristics follows clearly from his overall morphology. The numerous pithecoid traits that he has preserved are all reminders, still very vivid, of an ancestral state." And elsewhere (*ibid.*, 238): "The probable absence of all traces of preoccupations of an ethical or moral nature, accords well with the brutal appearance of this vigorous and heavy body, this bony head, these powerful jaws, which give evidence of the still existing prevalence of the purely vegetative or bestial functions over the cerebral functions." Boule did qualify his statements somewhat, saying (*ibid.*, 242–243): "Undoubtedly some of these traits are possibly not primitive, but rather the result of special physiological adaptations." Nevertheless he maintained that (*idem*): "There can be no doubt that most of them have a real genealogical significance." For an interesting analysis of the background to the 'Boulean' image of Neanderthal Man, see Hammond, "The Expulsion of the Neanderthals".

[92] Dubois, "On the Significance of the Large Cranial Capacity", 1921, 1271–1280.

[93] *Ibid.*, 1280–1288.

[94] Dubois, "On the Cranial Form", 1922, 315.

[95] *Ibid.*, 317–325.

[96] *Ibid.*, 331–332.

[97] Dubois, "Phylogenetic and Ontogenetic Increase", 1923, 231–232.

[98] Dubois, "On the Cranial Form", 1922, 332.

[99] Dubois, "Over de voornaamste onderscheidende eigenschappen", 1924, 148. As early as 1912 Dubois considered it less likely that Pithecanthropus belonged to a separate family ("Over de plaats van Pithecanthropus", 1912, 7). Compare with chapter 4, p. 119.

[100] *Ibid.*, 145.

[101] Dubois, "On the Brain Quantity", 1924, 435 note.

[102] In the English version of his description of the Pithecanthropus fossils he also mentioned this ratio ("On the Principal Characters", 1924, 274). In the Dutch version (see note 99) he had still been working with the ratio 3 : 2 : 1. It is worth mentioning that, in the English version, the observation that humans could not have descended directly from Pithecanthropus was toned down to the more neutral proposition that: "Man and Pithecanthropus... both descend from a common primitive Simian ancestor" (*idem*).

[103] Dubois, "On the Principal Characters of the Femur", 1926.

[104] Dubois, "Phylogenetic and Ontogenetic Increase", 1923, 232.

[105] Dubois, "On the Principal Characters of the Femur", 1926, 731.

[106] *Ibid.*, 731–732.

[107] *Ibid.*, 734–737.

[108] *Ibid.*, 737–739. The passage quoted is on p. 739.

[109] *Ibid.*, 739–741.

[110] *Ibid.*, 741.

[111] *Ibid.*, 742.

[112] *Ibid.*, 742–743.

[113] Dubois, "Die phylogenetische Grosshirnzunahme", 1930, 266–270.

[114] Dubois, "On the Gibbon-Like Appearance", 1935, 578–582.

[115] For this, see Brongersma, "Pithecanthropus, echt en 'pseudo'".

[116] Dubois, "The Distinct Organization", 1932; "New Evidence", 1934; "The Sixth (Fifth New) Femur", 1935; "The Osteone Arrangement", 1937. In 1934 Dubois suggested that the Pithecanthropus femur differed from human femora in its internal structure, as well. In his view these structural differences were linked with the difference in movement ("New Evidence", 1934). However, in 1937 he withdrew this argument. Closer study of the internal structure of human femora had made him see that the differences in the Pithecanthropus thigh bone he had assumed, had been based on incorrect observations ("The Osteone Arrangement", 1937).

[117] Dubois, "De afzonderlijke organisatie", 1932, 77.
[118] See for example Von Koenigswald, "Fossil Hominids", 59; Wendt, *Ich suchte Adam*; Eldredge and Tattersall, *The Myths*, 75; Reader, *Missing Links*, 52; Brace, "Tales", 414.
[119] Dubois, "On the Gibbon-Like Appearance", 1935, *passim*.
[120] Davidson Black, "On an Adolescent Skull".
[121] Weidenreich, "Sinanthropus pekinensis und seine Bedeutung".
[122] Weinert, "Der 'Sinanthropus pekinensis'".
[123] Boule, "Le Sinanthrope".
[124] Dubois, "The Shape and Size", 1933.
[125] *Ibid.*, 417–420.
[126] *Ibid.*, 420–422.
[127] *Ibid.*, 422.
[128] *Idem.*
[129] *Ibid.*, 423.
[130] Hrdlička, *The Skeletal Remains*, 365–368. Elliot Smith, "The Peking Man".
[131] Weidenreich, "The New Discoveries of Sinanthropus".
[132] Weidenreich, "The New Discovery of Three Skulls".
[133] Weidenreich, "The New Discoveries of Sinanthropus", 454.
[134] Dubois, "Racial Identity", 1936.
[135] Weidenreich, "Sinanthropus pekinensis and Its Position", 284.
[136] Dubois, "Racial Identity", 1936, 1180–1181.
[137] See Oppenoorth, "Homo (Javanthropus) soloensis".
[138] Dubois, "Racial Identity", 1936, 1182–1185. Dubois had already pointed to the similarity between Solo Man and the Wadjak fossils earlier on, in his "Early Man in Java", 1932, 20. For his ideas on Solo Man, see also "On the Fossil Human Skulls", 1937, 1–4.
[139] Dubois, "Racial Identity", 1185.
[140] For this see also Dubois, Review of *The Age of Pithecanthropus* by L.J.C. van Es, 1934.
[141] Weidenreich, "Discovery of the Femur".
[142] Dubois, "On the Fossil Human Skull", 1938, 386; "The Fossil Human Remains. Continuation", 1940, 850.
[143] For Von Koenigswald see: Franzen, "In Memoriam G.H.R. von Koenigswald"; Tobias, "The Life and Work".
[144] Von Koenigswald and Weidenreich, "The Relationship between Pithecanthropus and Sinanthropus", 928.
[145] Von Koenigswald's announcement of the find appeared under the title "Een nieuwe Pithecanthropus ontdekt". Dubois protested in an article entitled "Nieuwe Pithecanthropus ontdekt?", 1936. Von Koenigswald's official account bore the title "Erste Mitteilung über einen fossilen Hominiden".
[146] Dubois, "Racial Identity", 1936, 1181–1185.
[147] Weidenreich, "Pithecanthropus and Sinanthropus".
[148] Dubois, "The Mandible", 1938.
[149] Weidenreich, "Pithecanthropus and Sinanthropus".
[150] Von Koenigswald, "Ein Unterkieferfragment".
[151] Von Koenigswald, "Fossil Hominids".
[152] Von Koenigswald, "Ein neuer Pithecanthropus-Schädel".
[153] Dubois, "On the Fossil Human Skull", 1938, 384–385; "The Fossil Human Remains", 1940, 494.
[154] Von Koenigswald, "Ein neuer Pithecanthropus-Schädel", 185–186; "Pithecanthropus Received into the Human Family". For the history of the discovery, see Von Koenigswald, *Begegnungen mit dem Vormenschen*.
[155] Dubois, "On the Fossil Human Skull", 1938, 384–385.

[156] For this, see Von Koenigswald, *Begegnungen mit dem Vormenschen*. The protest was not published.
[157] Dubois, "Explanatory Statement", 1939.
[158] See for example Von Koenigswald, "De eerste 'aapmensen'", 248–249.
[159] Weidenreich, "Pithecanthropus and Sinanthropus".
[160] Von Koenigswald and Weidenreich, "The Relationship between Sinanthropus and Pithecanthropus".
[161] Dubois, "The Fossil Human Remains", 1940; "The Fossil Human Remains. Continuation", 1940; "The Fossil Human Remains. Conclusion", 1940.
[162] Dubois, "The Fossil Human Remains. Continuation", 1940, 848.
[163] Dubois, "The Fossil Human Remains", 1940, *passim*; "The Fossil Human Remains. Continuation", 1940, *passim*. To prevent any misunderstanding about his intentions, Dubois altered the name *Homo soloensis* to *Homo sapiens soloensis*.
[164] Dubois "The Fossil Human Remains. Continuation", 1940, 845.
[165] *Ibid.*, 844.
[166] Dubois, "The Fossil Human Remains. Conclusion", 1940, 1275.
[167] Quoted in Reader, *Missing Links*, 54.
[168] Brummelkamp, *Normale en abnormale hersengroei*; "Brainweight and Bodysize".
[169] For example see Versluys, *Hirngrösse und hormonales Geschehen*, 1–28; Bijlmer, *De evolutie van den mens*, 48–51; Droogleever Fortuyn, "The Cephalization Factor"; Bok, *Histonomy*, chapter 8; Van Erp Taalman Kip, *Lichaamsgrootte en hersenschors*, chapter 13. Bok only published the work mentioned above in 1959, but the research on which it was based was largely carried out in the 1930s. It is striking that Bok completely ignored the severe criticism aimed at the cephalisation theory in the 1930s and 1940s. The work by Van Erp Taalman Kip cited here was his doctoral thesis. He carried out the research for it under Bok's guidance. Van Erp was critical about several aspects of the cephalisation theory. He was unable to confirm either Dubois' or Brummelkamp's results in the matter of size relations between the constituent parts of nerve cells.
[170] See for example Ariëns Kappers, *The Evolution of the Nervous System*, 198–217.
[171] See Lapicque, "Sur le développement phylogénétique".
[172] Von Bonin, "Brain-Weight and Body-Weight".
[173] Anthony, "Essai de recherche".
[174] Sholl, "The Quantitative Investigation", 254.
[175] Count, "Brain and Body Weight"; Wirz, "Studien über die Cerebralisation".
[176] See Dubois, "Over de spronggrootte", 1938. According to Dubois a large number of the data on weight were worthless because they were obtained from animals living in captivity, which in general were lighter than average. He was at least partly right in this. For instance, Brummelkamp gave body weights of approximately 19 and 37 kilos for two adult chimpanzees, values that are indeed considerably below the average for these animals.
[177] The term 'allometry' was introduced in 1936 by Huxley and Teissier in their "Terminology of Relative Growth".
[178] Klatt, "Zur Methodik".
[179] Anthony and Coupin, "l'Indice de valeur cérébrale".
[180] Huxley became especially interested in the phenomenon of allometric growth after reading D'Arcy Wentworth Thompson's *On Growth and Form* of 1917 (see his *Problems of Relative Growth*, XVI–XVII). Thompson's work dealt with a spectrum of form and growth phenomena far beyond allometry alone, and also had a totally different background. Thompson's aim was especially to explain form by the direct effect of physical factors on the organism. Yet he also tried to find a mathematical formulation wherever possible. For more extensive information on this, see Gould, "D'Arcy Thompson and the Science of Form". Thompson was not particularly impressed by Dubois' studies of the brain:

"Dubois has attempted, but I think with very doubtful success, to equate the weight of the brain with that of the animal" (*On Growth and Form*, 92). In Thompson's view, Dubois' approach produced nothing more than empirical rules (*idem*). As far as I have been able to ascertain, Dubois never expressed any opinion of Thompson's work, and neither do we know his verdict on Huxley's *Problems of Relative Growth*. Neither book is in that part of his library incorporated into the Dubois Collection, so that it is not even certain that he knew these books.

181 See for example Jerison, *Evolution of the Brain*; further references can be found here.

182 Critical surveys can be found in, for example, White and Gould, "Interpretation of the Coefficient"; Gould, "Allometry and Size"; Holloway, "The Evolution of the Primate Brain"; Gould, "Geometric Similarity".

183 Allen, *Life Science*, XV.

184 *Ibid., XVII–XVIII.*

185 The ironical eye will observe that kindred spirits of Dubois have once more appeared in recent years. For example, there has been a recent attempt to lend support to a modern model of 'saltatory' evolution, to wit Gould's and Eldredge's 'punctuated equilibria' theory, by means of a study of cephalisation in fossil mammals and hominids. I hasten to emphasise that in no way is this a direct continuation of Dubois' work. I only wish to register the fact that Dubois' research method has maintained its value in the study of evolution. See Hofman, "Encephalization in Mammals"; "Encephalization in Hominids". Hofman increases the 'phylogenetic exponent' to 0.732, where for years it was given a value of approximately ⅔.

186 Dubois, "Pithecanthropus erectus als organisme", 1935, 156.

187 Dubois to J.H.F. Kohlbrugge, 6 April 1920.

188 Starck, "Die idealistische Morphologie", 52–54. See also Dullemeijer, *Concepts and Approaches*, 4–6, *passim.*

189 A functional anatomist of the Van der Klaauw school is P. Dullemeijer, for instance. See his *Concepts and Approaches*, in which further references may be found.

190 Zacharias, *The Construction of a Primate Order*, chapter 8.

191 *Ibid.*, 379–395. Here Zacharias also discusses the way in which Keith made use of the available fossil material, such as the Pithecanthropus remains, in his theoretical considerations.

192 Zacharias (*ibid.*, 362–363) has interpreted the shift in attention among English morphologists as confirmation of Allen's thesis that there was a revolt against morphology around 1900 (Allen, *Life Science*, 21–39). In my view she has misinterpreted Allen's intentions here. Allen used the term 'morphology' not merely to signify classic morphology, oriented towards aspects of form, but as an umbrella term for the descriptive, non-quantitative disciplines within biology. He saw the revolt as resulting, moreover, in a transfer of interest to the experimental disciplines. According to Allen a pronounced dichotomy appeared between descriptive and experimental work, between 'naturalists' and 'experimentalists' ("Morphology and Twentieth-Century Biology"). Therefore the development of a functional approach within morphology reveals a weakness in Allen's model, since it provides no place for such developments. Also, Dubois' research on cephalisation, purely morphological in nature, but quantitative at the same time, vanishes into a no-man's-land because of Allen's naturalist-experimentalist dichotomy. For a number of critical discussions of Allen's model, see: "Special Section on American Morphology".

193 Dubois to J.H.F. Kohlbrugge, 6 April 1920.

194 Dubois, "Eenige uitkomsten der palaeontologie", 1927, 151.

195 Dubois to Jos. Cremers, 1 March 1940 (*Natuurhistorisch Museum*, Maastricht).

196 See note 80.

197 See for example Allen, *Life Science*, 73–126, especially 103–106.

[198] For Jordan's ideas on the matter, see Visser, "De Utrechtse zoölogie", 55–59.

[199] Dubois, "Die phylogenetische Grosshirnzunahme", 1930, 286. It goes without saying that Dubois' ideas on the development of life were incompatible with those of the British and American biologists who brought about the 'evolutionary synthesis' in the 1930s and 1940s (see Mayr and Provine, *The Evolutionary Synthesis*). All that can be gathered from Dubois' publications and his unpublished papers and correspondence, is that he either had not kept up with developments in Anglo-American biology, or that he completely ignored them. For Dubois Darwinism was finished, and there is nothing to indicate that he was ever concerned with the relationship between genetics and evolution. The whole subject of genetics is conspicuous by its absence in his work. In a letter to G.A.F. Molengraaff on 24 June 1930, Dubois wrote that geneticists were "basically botanists who study artificial selection." This remark arouses the strong impression that Dubois was completely unaware of contemporary developments in genetics.

[200] See Driesch, *Analytische Theorie*. Driesch wrote for instance (*ibid*, 137–138) that: "Every breed, in the Darwinian sense, is supposed to have developed by gradual improvement; many imperfect intermediate stages together bring about the maturation of the perfect in the course of the generations. How unfortunate, though, that ontogenetic mechanisms cannot bring a full-blown organism into existence, unless they function perfectly and completely all the time! Any 'transition', any 'advance', is logically excluded here!" See also Churchill, "From Machine-Theory to Entelechy", 175.

[201] Dubois, "The Fossil Human Remains. Continuation", 1940, 851.

SOURCES

Manuscript sources

For this study comprehensive use has been made of the Dubois papers in the *Rijksmuseum van Natuurlijke Historie* in Leiden, which together with the vertebrate fossils Dubois excavated in the Dutch East Indies constitute the Dubois Collection. These papers include manuscripts, correspondence, lecture notes, annotations, diaries, and so on. The Dubois Collection also contains Dubois' palaeontological and palaeoanthropological library, an extensive collection of reprints, and photographs. The records are not catalogued. The correspondence is arranged alphabetically, while the remaining records are classified according to subject.

Unless otherwise indicated, the manuscript sources cited in this study are held in the Dubois Collection. Where the source of letters written by Dubois is not mentioned, I have used copies, transcripts, or outlines thereof kept in the Dubois Collection.

A supplementary source is the four-part diary kept by the palaeontologist J.J.A. Bernsen O.F.M. From 1930 to 1932 Bernsen assisted Dubois in his work on the fossil collection, and in his diary he gave a detailed account of his conversations with Dubois in this period. Mrs. C.J.M. van Melsen–Bernsen allowed me to use a typescript of her uncle's diary, and gave permission for publication of the passages quoted. Bernsen's diary itself is in the archives of the Franciscan provincialate at Utrecht.

Publications by Dubois

This is not a full bibliography of Dubois' writings. Only the publications cited in this book and all translations of these works have been included. The paragraph headings indicate the year of publication. They correspond to the years in the notes. The following abbreviations are used:

N.T.: *Natuurkundig tijdschrift voor Nederlandsch-Indië*
Proceedings: *Proceedings of the Section of Sciences of the Koninklijke Akademie van Wetenschappen*
T.A.G.: *Tijdschrift van het Koninklijk Nederlandsch Aardrijkskundig Genootschap*

Verslag: *Verslag van de gewone vergaderingen der wis- en natuurkundige afdeeling der Koninklijke Akademie van Wetenschappen te Amsterdam.* From vol. 34 (1925) continued as: *Verslag van de gewone vergaderingen der afdeeling natuurkunde, Koninklijke Akademie van Wetenschappen te Amsterdam*

V.M.: *Verslag van het Mijnwezen. Extra bijvoegsel der Javasche courant*

1886
"Ueber den Larynx", in: M. Weber, *Studien über Säugethiere: ein Beitrag zur Frage nach dem Ursprung der Cetaceen* (Jena, 1886).
"Zur Morphologie des Larynx", *Anatomischer Anzeiger* 1 (1886) 178–186, 225–231.

1888
"Over de wenschelijkheid van een onderzoek naar de diluviale fauna van Ned. Indië, in het bijzonder van Sumatra", *N.T.* 48 (1889) 148–165.

1889
[Uit een schrijven van den heer Dubois te Pajacombo naar aanleiding van den aan dien heer toegezonden schedel, door den heer van Rietschoten in zijn marmergroeven in het Kedirische opgegraven], *N.T.* 49 (1890) 209–210.

1891
"Palaeontologische onderzoekingen op Java", *V.M.* (3rd quarter 1890) nr. 9 (1891) 12–15.
"Palaeontologische onderzoekingen op Java", *V.M.* (4th quarter 1890) nr. 41 (1891) 14–18.
"Voorloopig bericht omtrent het onderzoek naar de pleistocene en tertiaire vertebraten-fauna van Sumatra en Java, gedurende het jaar 1890", *N.T.* 51 (1892) 93–100.
"Palaeontologische onderzoekingen op Java", *V.M.* (1st quarter 1891) nr. 58 (1891) 12–13.
"Palaeontologische onderzoekingen op Java", *V.M.* (2nd quarter 1891) nr. 93 (1891) 11–12.

1892
"Palaeontologische onderzoekingen op Java", *V.M.* (3rd quarter 1891) nr. 3 (1892) 12–14.
"Palaeontologische onderzoekingen op Java", *V.M.* (4th quarter 1891) nr. 36 (1892) 12–15.
"Palaeontologische onderzoekingen op Java", *V.M.* (2nd quarter 1892) nr. 87 (1892) 14–18.

1893
"Palaeontologische onderzoekingen op Java", *V.M.* (3rd quarter 1892) nr. 10 (1893) 10–14.
"Palaeontologische onderzoekingen op Java", *V.M.* (4th quarter 1892) nr. 36 (1893) 11–12.

1894
"Palaeontologische onderzoekingen op Java", *V.M.* (3rd quarter 1893) nr. 2 (1894) 15–17.
"Palaeontologische onderzoekingen op Java", *V.M.* (2nd quarter 1894) nr. 81 (1894) 12–15.
Pithecanthropus erectus, eine menschenähnliche Uebergangsform aus Java (Batavia, 1894).

1895
"Pithecanthropus erectus, betrachtet als eine wirkliche Uebergangsform und als Stammform des Menschen", *Zeitschrift für Ethnologie* 27 (1895) 723–738.
"Pithecanthropus erectus, eine menschenähnliche Uebergangsform aus Java", *Jaarboek van het Mijnwezen in Nederlandsch-Oost-Indië* 24 (1895) 1–77.
"Sur le Pithecanthropus erectus du pliocène de Java. Résumé", *Bulletin de la Société Belge de Géologie, de Paléontologie et d'Hydrologie. Proces-verbaux* 9 (1895) 151–160.
"The Place of 'Pithecanthropus' in the Genealogical Tree", *Nature* 53 (1895–1896) 245, 247.

1896
"Näheres über den Pithecanthropus erectus als menschenähnliche Uebergangsform", *Internationale Monatschrift für Anatomie und Physiologie* 13 (1896) 1–26.
"On Pithecanthropus erectus: a Transitional Form between Man and the Apes", *Scientific Transactions of the Royal Dublin Society* S. 2, 6 (1896) 1–18.
"On Pithecanthropus erectus: a Transitional Form between Man and the Apes. Abstract", *Journal of the Anthropological Institute of Great-Britain and Ireland* 25 (1896) 240–248.
"Pithecanthropus erectus, eine menschenähnliche Uebergangsform", in: *Compte-rendu des séances du Troisième Congrès International de Zoologie, Leyde, 11–16 septembre 1895* (Leiden, 1896) 251–271.
"Pithecanthropus erectus, eine Stammform des Menschen", *Anatomischer Anzeiger* 12 (1896) 1–22.
"De thans bekende soorten van fossiele menschapen", *Tijdschrift der Nederlandsche Dierkundige Vereeniging* S. 2, 5 (1896) LXX–LXXIV.
"Le Pithecanthropus erectus et l'origine de l'homme", *Bulletins de la Société d'Anthropologie de Paris* S. 4, 7 (1896) 460–467.

1897
"De verhouding van het gewicht der hersenen tot de grootte van het lichaam bij de zoogdieren", *Verhandelingen der Koninklijke Akademie van Wetenschappen* S. 2, 5, nr. 10 (1897) 1–41.
Ueber drei ausgestorbene Menschenaffen", *Neues Jahrbuch für Mineralogie, Geologie und Palaeontologie* (1897) 83–104.
"Sur le rapport du poids de l'encéphale avec la grandeur du corps chez les mammifères", *Bulletins de la Société d'Anthropologie de Paris* S. 4, 8 (1897) 337–376.

1898
Ueber die Abhängigkeit des Hirngewichtes von der Körpergrösse bei den Säugethieren", *Archiv für Anthropologie* 25 (1898) 1–28.
Ueber die Abhängigkeit des Hirngewichtes von der Körpergrösse beim Menschen", *Archiv für Anthropologie* 25 (1898) 423–441.

1899
"Pithecanthropus erectus: a Form from the Ancestral Stock of Mankind", *Report of the Smithsonian Institution for 1898* (Washington, 1899) 445–449.
"Remarks upon the Brain–Cast of Pithecanthropus erectus", in: *Proceedings of the Fourth International Congres of Zoology, Cambridge, 22–27 August 1898* (London, 1899) 78–95.

1901
"Zur systematischen Stellung der ausgestorbenen Menschenaffen", *Zoologischer Anzeiger* 24 (1901) 556–560.

1902
"Données justificatives sur l'essai de réconstruction plastique du Pithecanthropus erectus", *Petrus Camper. Nederlandsche bijdragen tot de anatomie* 1 (1902) 237–241.

1907
"Eenige van Nederlandschen kant verkregen uitkomsten met betrekking tot de kennis der Kendeng-fauna (fauna van Trinil)", *T.A.G.* S. 2, 24 (1907) 449–458.

1908
"Das geologische Alter der Kendeng- oder Trinil-Fauna", *T.A.G.* S. 2, 25 (1908) 1235–1270.

1909
[Die fossile Zähne von Trinil], *T.A.G.* S. 2, 26 (1909) 399–401.

1912
"Over de plaats van Pithecanthropus in het zoölogisch systeem", *Archives du Musée Teyler* S. 3, 1 (1912) 142–149.

1914
"De betrekking tusschen hersenmassa en lichaamsgrootte bij de gewervelde dieren", *Verslag* 22 (1913–1914) 593–614.
"Die gesetzmässige Beziehung von Gehirnmasse zu Körpergrösse bei den Wirbeltieren", *Zeitschrift für Morphologie und Anthropologie* 18 (1914) 323–350.
"On the Relation between the Quantity of Brain and the Size of the Body in Vertebrates", *Proceedings* 16 (1914) 647–668.

1918
"De betrekking der hoeveelheden van de hersenen, het neuron en zijn deelen tot de lichaamsgrootte", *Verslag* 26 (1917–1918) 1416–1425.
"On the Relation between the Quantities of the Brain, the Neurone and its Parts, and the Size of the Body", *Proceedings* 20 (1918) 1328–1337.

1919
"De beteekenis der grootte van het neuron en zijn deelen", *Verslag* 27 (1918–1919) 503–520.
"Vergelijking van het hersengewicht, in functie van het lichaamsgewicht, tusschen de twee seksen", *Verslag* 27 (1918–1919) 713–732.
"The Significance of the Size of the Neurone and its Parts", *Proceedings* 21 (1919) 711–729.
"Comparison of the Brain Weight in Function of the Body Weight, between the Two Sexes", *Proceedings* 21 (1919) 850–869.

1920
"De hoeveelheidsbetrekkingen van het zenuwstelsel bepaald door het mechanisme van het neuron", *Verslag* 28 (1919–1920) 623–638.
"De hoeveelheidsbetrekkingen van de hersenen tot het lichaam der gewervelde diersoorten en hare beteekenis", *Vakblad voor biologen* 1 (1920) 82–89.
"The Quantitative Relations of the Nervous System Determined by the Mechanism of the Neurone", *Proceedings* 22 (1920) 665–680.

1921
"De proto-Australische fossiele mensch van Wadjak, Java", *Verslag* 29 (1920–1921) 88–105.
"De proto-Australische fossiele mensch van Wadjak, Java. II", *Verslag* 29 (1920–1921) 866–887.
"De beteekenis der groote schedelcapaciteit van Homo neandertalensis", *Verslag* 29 (1920–1921) 987–1004.

"Over den schedelvorm van Homo neandertalensis en van Pithecanthropus erectus, bepaald door mechanische factoren", *Verslag* 30 (1921) 391–411.
"The Proto-Australian Fossil Man of Wadjak, Java", *Proceedings* 23 (1921) 1013–1051.
"On the Significance of the Large Cranial Capacity of Homo neandertalensis", *Proceedings* 23 (1921) 1271–1288.

1922
"Hat sich das Gehirn beim Haushunde, im Vergleich mit Wildhundarten, vergrössert, oder verkleinert?", *Bijdragen tot de dierkunde* 22 (1922) 315–320.
"On the Cranial Form of Homo neandertalensis and of Pithecanthropus erectus, Determined by Mechanical Factors", *Proceedings* 24 (1922) 313–332.
"Phylogenetische en ontogenetische toeneming van het volumen der hersenen bij de gewervelde dieren", *Verslag* 31 (1922) 307–337.

1923
"Phylogenetic and Ontogenetic Increase of the Volume of the Brain in Vertebrata", *Proceedings* 25 (1923) 230–255.

1924
"Over de voornaamste onderscheidende eigenschappen van den schedel en de hersenen, de onderkaak en het gebit van Pithecanthropus erectus", *Verslag* 33 (1924) 135–148.
"On the Principal Characters of the Cranium and the Brain, the Mandible and the Teeth of Pithecanthropus erectus", *Proceedings* 27 (1924) 265–278.
"Figures of the Calvarium and Endocranial Cast, a Fragment of the Mandible and Three Teeth of Pithecanthropus erectus", *Proceedings* 27 (1924) 459–464.
"Over de hersenhoeveelheid van gespecialiseerde zoogdiergeslachten", *Verslag* 33 (1924) 319–326.
"On the Brain Quantity of Specialized Genera of Mammals", *Proceedings* 27 (1924) 430–437.

1925
"Over de plaats van den Pithecanthropus onder de primaten en zijn genealogische betrekkingen, naar de beschouwingen van Boule", *Vakblad voor biologen* 7 (1925) 1–5.

1926
"Over de voornaamste onderscheidende eigenschappen van het femur van Pithecanthropus erectus", *Verslag* 35 (1926) 443–455.

"On the Principal Characters of the Femur of Pithecanthropus erectus", *Proceedings* 29 (1926) 730–743.
"Figures of the Femur of Pithecanthropus erectus", *Proceedings* 29 (1926) 1275–1277.
"Manis palaejavanica, het reuzenschubdier der Kendeng-fauna", *Verslag* 35 (1926) 949–958.
"Manis palaejavanica, the Giant Pangolin of the Kendeng Fauna", *Proceedings* 29 (1926) 1233–1243.

1927
"Opmerkingen over eenige uitkomsten der palaeontologie met betrekking tot de evolutie der dierenwereld", in: *Handelingen van het 21ste Nederlandsch Natuur- en Geneeskundig Congres, Amsterdam, 19–21 april 1927* (Amsterdam, 1927) 150–151.
"Ueber die Hauptmerkmale des Femur von Pithecanthropus erectus", *Anthropologischer Anzeiger* 4 (1927) 131–146.

1928
"De wet der noodwendige phylogenetische volmaking van het psychoencephalon", *Verslag* 37 (1928) 252–259.
"The Law of the Necessary Phylogenetic Perfection of the Psychoencephalon", *Proceedings* 31 (1928) 304–314.

1930
"Phyloblastese, het beginsel der phylogenese en de grondvoorwaarde der aanpassingsbetrekkingen", *Verslag* 39 (1930) 124–127.
"Die phylogenetische Grosshirnzunahme autonome Vervollkommnung der animalen Funktionen", *Biologia Generalis* 6 (1930) 247–292.

1932
"De afzonderlijke organisatie van Pithecanthropus, waarvan het femur getuigt, thans bevestigd door andere individuen van de beschreven soort", *Verslag* 41 (1932) 76–77.
The Distinct Organization of Pithecanthropus of Which the Femur Bears Evidence, Now Confirmed from Other Individuals of the Described Species", *Proceedings* 35 (1932) 716–722.
"Early Man in Java", *Nature* 130 (1932) 20.

1933
"The Seeming and the Real Cephalization of the Australian Aborigine", *Proceedings* 36 (1933) 2–13.
"De vorm en de grootte der hersenen bij Sinanthropus en bij Pithecanthropus", *Verslag* 42 (1933) 40–42.

"The Shape and the Size of the Brain in Sinanthropus and in Pithecanthropus", *Proceedings* 36 (1933) 415–423.

1934

"New Evidence of the Distinct Organization of Pithecanthropus", *Proceedings* 37 (1934) 139–145.

"Phylogenetic Cerebral Growth", in: *Comptes rendus du Premier Congres International des Sciences Anthropologiques et Ethnologiques, Londres 1934* (London, 1934) 71–75.

"Ueber die Gleichheit der Cephalisationsstufe aller Menschengruppen bei verschiedenen spezifischen Gehirnvolumen", *Biologia Generalis* 10 (1934) 185–193.

Review of *The Age of Pithecanthropus* (Den Haag, 1931) by L.J.C. van Es, *Mensch en maatschappij* 10 nr. 3 (1934) 216–221.

1935

"On the Gibbon-Like Appearance of Pithecanthropus erectus", *Proceedings* 38 (1935) 578–585.

"Pithecanthropus erectus als organisme", in: *Handelingen van het 25ste Nederlandsch Natuur- en Geneeskundig Congres, Leiden, 23–25 april 1935* (Leiden, 1935) 147–159.

"The Sixth (Fifth New) Femur of Pithecanthropus erectus", *Proceedings* 38 (1935) 850–852.

1936

"Nieuwe Pithecanthropus ontdekt?", *Algemeen handelsblad*, evening-paper 18 April 1936.

"Racial Identity of Homo soloensis Oppenoorth (Including Homo modjokertensis Von Koeningswald) and Sinanthropus pekinensis Davidson Black", *Proceedings* 39 (1936) 1180–1185.

1937

"Early Man in Java and Pithecanthropus erectus", in: G.G. MacCurdy, ed. *Early Man as Depicted by Leading Authorities at the International Symposium. The Academy of Natural Sciences, Philadelphia, March 1937* (Philadelphia, 1937) 315–322.

"On the Fossil Human Skulls Recently Discovered in Java and Pithecanthropus erectus", *Man* 37 (1937) 1–7.

"The Osteone Arrangement of the Thigh-Bone Compacta of Man Identical with That, First Found of Pithecanthropus", *Proceedings* 40 (1937(864–870.

1938

"The Mandible Recently Described and Attributed to the Pithecanthropus by G.H.R. von Koenigswald, Compared with the Mandible of Pithecan-

thropus erectus Described in 1924 by Eug. Dubois", *Proceedings* 41 (1938) 139–147.
"On the Fossil Human Skull Recently Described and Attributed to Pithecanthropus erectus by G.H.R. von Koenigswald", *Proceedings* 41 (1938) 380–386.
"Over de spronggrootte van den phylogenetischen en ontogenetischen hersengroei", *Vakblad voor biologen* 20 (1938) 37–46.

1939
"Explanatory Statement Concerning the Communication of Eug. Dubois in Proceedings K.N.A.v.W., vol. XLI, no. 4, 1938: On the Fossil Human Skull Recently Described and Attributed to Pithecanthropus erectus by G.H.R. von Koenigswald", *Proceedings* 42 (1939) 53.
"On Insignificance of Cranial Vault-Height in Phylogenetic Braingrowth", *Proceedings* 42 (1939) 125–126.

1940
"The Fossil Human Remains Discovered in Java by Dr. G.H.R. von Koenigswald and Attributed by Him to Pithecanthropus erectus, in Reality Remains of Homo wadjakensis (Syn. Homo soloensis)", *Proceedings* 43 (1940) 494–496.
"The Fossil Human Remains Discovered in Java by G.H.R. von Koenigswald and Attributed by Him to Pithecanthropus erectus, in Reality Remains of Homo sapiens soloensis. Continuation", *Proceedings* 43 (1940) 842–851.
"The Fossil Human Remains Discovered in Java by G.H.R. von Koenigswald and Attributed by Him to Pithecanthropus erectus, in Reality Remains of Homo sapiens soloensis. Conclusion", *Proceedings* 43 (1940) 1268–1275.

Literature

Ackerknecht, E.H. *Rudolf Virchow: Arzt, Politiker, Anthropologe* (Stuttgart, 1957).

Allen, G.E. *Life Science in the Twenthieth Century* (New York, 1975).

Allen, G.E. "Morphology and Twentieth-Century Biology: a Response", *Journal of the History of Biology* 14 (1981) 159–176.

Andrews, R.C. *On the Trail of Ancient Man: a Narrative of the Fieldwork of the Central Asiatic Expeditions* (New York, 1926).

Anonymus. "Belangrijke palaeontologische vondsten op Java", *Tijdschrift van het Koninklijk Nederlandsch Aardrijkskundig Genootschap* S. 2, 10 (1893) 310–312.

Anonymus. "The Third International Zoological Congress at Leyden", *Nature* 52 (1895) 554–555.

Anonymus. "Dr. Dubois' 'Missing Link'", *Nature* 53 (1895–1896) 115–116.

Anonymus. Report on a meeting at the Anthropological Institute, *Journal of the Anthropological Institute* 96 (1896) 248–255.

Anonymus. "Prof. Dubois blikt over zijn leven terug", *Telegraaf*, 27 January 1938.

Anthony, R. "Essai de recherche d'une expression anatomique approximative du degré d'organisation cérébrale autre que le poids de l'encéphale comparé au poids du corps", *Bulletins et mémoires de la Société d'Anthropologie* S. 9, 8 (1938) 17–67.

Anthony, R. and Coupin, F. "l'Indice de valeur cérébrale", *Revue anthropologique* 35 (1925) 145–151.

Ariëns Kappers, C.U. *The Evolution of the Nervous System in Invertebrates, Vertebrates and Man* (Haarlem, 1929).

Barge, J.A.J. "In memoriam Eugène Dubois", *Nederlandsch tijdschrift voor geneeskunde* 84 (1940) 4994–4995.

Bartholomew, M. "Lyell and Evolution: an Account of Lyell's Response to the Prospect of an Evolutionary Ancestry for Man", *British Journal for the History of Science* 6 (1973) 261–303.

Bartholomew, M. "Huxley's Defense of Darwin", *Annals of Science* 32 (1975) 525–535.

B[emmelen], [J.F.] v[an]. "Het Leidsche internationale zoölogencongres: Dubois' aapmensch voor de vierschaar der wetenschap", *Java-bode*, 16 November 1895.

Bemmelen, J.M. van. "Der Gehalt an Fluorcalcium eines fossilen Elephantenknochen aus der Tertiärzeit", *Zeitschrift für anorganische Chemie* 15 (1897) 84–122.

Bemmelen, J.M. van. "Sur la teneur en fluoride de calcium d'un os d'éléphant fossile de l'époque tertiaire", *Archives néerlandaises des sciences exactes et naturelles* S. 2, 3 (1900) 229–235.

Bernsen, J.J.A. *The Geology of the Teglian Clay and Its Fossil Remains of Rhinoceros* ('s-Hertogenbosch, 1927).

Bijlmer, H.J.T. *De evolutie van den mens* (Den Haag, 1946).

Bischoff, T.L.W. *Ueber die Verschiedenheit in der Schädelbildung des Gorilla, Chimpanse und Orang-Outang, vorzüglich nach Geschlecht und Alter, nebst einer Bemerkung über die Darwinische Theorie* (München, 1867).

Black, D. "On an Adolescent Skull of Sinanthropus pekinensis in Comparison with an Adult Skull of the Same Species and with Other Hominid Skulls Recent and Fossil", *Palaeontologia Sinica* Series D, 7, nr. 2 (1931) 1–144.

Bluntschli, H. "Max Fürbringer (1846–1920)", *Anatomischer Anzeiger* 55 (1922) 244–255.

Boas, F. "The Aims of Anthropological Research", *Science* N.S. 76 (1932) 605–613.

Boeke, J. "Levensbericht van Marie Eugène François Thomas Dubois (28 januari 1858 – 16 december 1940)", *Jaarboek der Nederlandsche Akademie van Wetenschappen* (1940–1941) 238–247.

Bok, S.T. *Histonomy of the Cerebral Cortex* (Amsterdam, 1959).

Bonin, G. von. "Brain-Weight and Body-Weight of Mammals", *Journal of General Psychology* 16 (1937) 379–389.

Bork-Feltkamp, A.J. van. "Anthropological Research in the Netherlands", *Verhandelingen der Koninklijke Nederlandsche Akademie van Wetenschappen, afdeeling natuurkunde* S. 2, 37, nr. 3 (1938) 1–166.

Bosch, M. van den. "J.G.S. van Breda en de Commissie voor de Geologische Kaart van Nederland, 1852–1855", in: A.S.H. Breure and J.G. de Bruijn, eds. *Leven en werken van J.G.S. van Breda (1788–1867)* (Haarlem, 1979) 267–402.

Boucher de Perthes, J. *Antiquités Celtiques et antédiluviennes: mémoire sur l'industrie primitive et les arts à leur origine* (Paris, 1847).

Boule, M. "l'Homme fossile de la Chapelle-aux-Saints", *Annales de paléontologie* 6 (1911) 1–64; 7 (1912) 65–208; 8 (1913) 209–279.

Boule, M. *Les hommes fossiles: éléments de paléontologie humaine* (Paris, 1921).

Boule, M. "Le Sinanthrope", *l'Anthropologie* 39 (1929) 455–460.

Bowler, P.J. *The Eclipse of Darwinism: Anti-Darwinian Evolution Theories in the Decades around 1900* (Baltimore, 1984).

Bowler, P.J. *Theories of Human Evolution: A Century of Debate, 1844–1944* (Baltimore, 1986).

Brace, C.L. "Tales of the Phylogenetic Woods: the Evolution and Significance of Evolutionary Trees", *American Journal of Physical Anthropology* 56 (1981) 411–429.

Branco, W. "Die menschenähnliche Zähne aus dem Bohnerz der schwäbischen Alb", *Jahreshefte des Vereins für Vaterländische Naturkunde in Württemberg* 54 (1898) 1–144.

Branco, W. "Vorläufiger Bericht über die Ergebnisse der Trinil-Expedition der Akademische Jubiläums-Stiftung der Stadt Berlin", *Sitzungsberichte der Königlichen Preussischen Akademie der Wissenschaften* (1908) 261–273.

Brom, G.B. *Herleving van de wetenschap in katholiek Nederland* ('s-Gravenhage, 1930).

Brongers, J.A. "Material for a History of Dutch Archaeology up to 1922", *Berichten van de Rijksdienst voor het Oudheidkundig Bodemonderzoek* 26 (1976) 7–62.

Brongersma, L.D. "De verzameling van Indische fossielen (Collectie Dubois)", *De Indische gids* 63 (1941) 97–116.

Brongersma, L.D. "Eugène Dubois", *Natuurhistorisch maandblad* 62 (1973) 107–109.

Brongersma, L.D. "Rijksmuseum van Geologie en Mineralogie, 1878–1978: Past, Present and Future", *Scripta Geologica* 48 (1978) 37–96.

Brongersma, L.D. "Pithecanthropus: echt en 'pseudo'", *Verslag van de gewone vergadering van de afdeling natuurkunde der Koninklijke Nederlandse Akademie van Wetenschappen* 91 (1982) 34–36.

Brongersma, L.D. "Een geologisch verleden", in: *GSVA bulletin, special edition* (Amsterdam, 1984) 3–12.

Brugmans, H., Scholte, J.H. and Kleintjes, Ph. *Gedenkboek van het Atheneum en de Universiteit van Amsterdam, 1632–1932* (Amsterdam, 1932).

Brummelkamp, R. *Normale en abnormale hersengroei in verband met de cephalisatie-leer* (Amsterdam, 1937).

Brummelkamp, R. "Brainweight and Bodysize: a Study of the Cephalization-Problem", *Verhandelingen der Koninklijke Nederlandsche Akademie van Wetenschappen, afdeeling natuurkunde* S. 2, 39, nr. 5 (1940) 1–57.

Buffon, G.L.L. de. *Histoire naturelle, générale et particulière avec la description du Cabinet du Roi* (36 vols; Paris, 1749–1788).

Bumüller, J. "Menschen- und Affen-Femur", *Mittheilungen der Anthropologischen Gesellschaft in Wien* 30 (1900) 91–94.

Burrow, J.W. "Evolution and Anthropology in the 1860's: the Anthropological Society of London, 1863–1871", *Victorian Studies* 7 (1963) 137–154.

Bynum, W.F. "Charles Lyell's Antiquity of Man and its Critics", *Journal of the History of Biology* 17 (1984) 153–187.

Calwer, C.G. *Käferbuch: allgemeine und specielle Naturgeschichte der Käfer Europa's* (Stuttgart, 1858).

Churchill, F.B. "From Machine-Theory to Entelechy: Two Studies in Developmental Teleology", *Journal of the History of Biology* 2 (1969) 165–185.

Coleman, W. *Biology in the Nineteenth Century: Problems of Form, Function, and Transformation* (New York, 1971).

Coleman, W. "Morphology between Type Concept and Descent Theory", *Journal of the History of Medicine and Allied Sciences* 31 (1976) 149–165.

Count, E.W. "Brain and Body Weight in Man: Their Antecedents in Growth and Evolution", *Annals of the New York Academy of Science* 46 (1947) 993–1122.

Cremers, J. "Excursie Rijckholt – St. Geertruid", *Natuurhistorisch maandblad* 12 (1923) 23–28.

Cunningham, D.J. "Dr. Dubois' So-Called Missing Link", *Nature* 51 (1894–1895) 428–429.

Cunningham, D.J. [A Paper on Pithecanthropus erectus, the Man-like Transitional Form of Dr. Eug. Dubois], *Journal of Anatomy and Physiology* 29, N.S. 9 (1895) XVIII–XIX.

Cunningham, D.J. "The Place of 'Pithecanthropus' on the Genealogical Tree", *Nature* 53 (1895–1896) 269.

Cuvier, G. *Discours sur les révolutions de la surface du globe, et sur les changemens qu'elles ont produits dans le règne animal* (Bruxelles, 1969; reprint of the third edition; Paris, 1825).

Dames, W. "Pithecanthropus: ein Bindeglied zwischen Affe und Mensch", *Deutsche Rundschau* 88 (1896) 368–384.

Darwin, Ch. *The Descent of Man, and Selection in Relation to Sex* (Princeton, New Jersey, 1981; reprint of the first edition; London, 1871).

Day, M.H. and Molleson, T.I. "The Trinil Femora", in: M.H. Day, ed. *Human Evolution* (London, 1973) 127–154.

Desmond, A.J. *Archetypes and Ancestors: Palaeontology in Victorian London, 1850–1875* (London, 1982).

Deyssel, L. van. *Blank en Geel* (Amsterdam, 1979; reissue of the first edition (1892), with an afterword by H.G.M. Prick).

Driesch, H. *Analytische Theorie der organischen Entwicklung* (Leipzig, 1894).

Droogleever Fortuyn, A.B. "The Cephalization Factor in the Hamster and the Mouse", *Psychiatrische en neurologische bladen* 38 (1934) 392–395.

Dullemeijer, P. *Concepts and Approaches in Animal Morphology* (Assen, 1974).

Eiseley, L.C. *Darwin's Century: Evolution and the Men Who Discovered It* (London, 1959).

Elbert, J. "De nieuwste onderzoekingen over het Pithecanthropusvraagstuk", *Natuurkundig tijdschrift voor Nederlandsch-Indië* 67 (1907) 125–142.

Eldredge, N. and Tattersall, I. *The Myths of Human Evolution* (New York, 1982).

Erickson, P.A. *The Origins of Physical Anthropology* (Diss. Univ. of Connecticut, 1974; Ann Arbor, Mich., 1976).

Erp Taalman Kip, M.J. van. *Lichaamsgrootte en hersenschors* (Groningen, 1938).

Farber, P.L. "The Type-Concept in Zoology during the First Half of the Nineteenth Century", *Journal of the History of Biology* 9 (1976) 93–119.

Farley, J. "The Initial Reactions of French Biologists to Darwin's 'Origin'", *Journal of the History of Biology* 7 (1974) 275–300.

Feen, P.J. van der and Benthem Jutting, W.S.S. van. "Antje Schreuder, Amsterdam, 15 november 1887 – Amsterdam, 2 februari 1952", *Geologie en mijnbouw* 14 (1952) 121–125.

Fischer, E. "Gustav Schwalbe", *Zeitschrift für Morphologie und Anthropologie* 20 (1917) I–VIII.

Flower, W.H. and Lydekker, R. *An Introduction to the Study of Mammals Living and Extinct* (London, 1891).

Fraipont, J. and Lohest, M. "La race humaine de Néanderthal ou de Canstadt en Belgique", *Archives de biologie* 7 (1886) 587–755.

Franzen, J.L. "In Memoriam G.H.R. von Koenigswald 1902–1982", *Senckenbergiana Lethaea* 64 (1983) 381–402.

Fürbringer, M. *Beitrag zur Kenntniss der Kehlkopfmuskulatur* (Jena, 1875).

Fürbringer, M. *Untersuchungen zur Morphologie und Systematik der Vögel, zugleich ein Beitrag zur Anatomie der Stütz- und Bewegungsorgane* (Amsterdam, 1888).

Gaudry, J.A. *Les enchaînements du monde animal dans les temps géologiques: mammifères tertiaires* (Paris, 1878).
Gaudry, J.A. "Le Dryopithèque", *Mémoires de la Société Géologique de France. Paléontologie* 1 (1890) 4–11.
Gould, S.J. "Allometry and Size in Ontogeny and Phylogeny", *Biological Reviews of the Cambridge Philosophical Society* 41 (1966) 587–640.
Gould, S.J. "D'Arcy Thompson and the Science of Form", *New Literary History* 2 (1971) 229–258.
Gould, S.J. "Geometric Similarity in Allometric Growth: a Contribution to the Problem of Scaling in the Evolution of Size", *American Naturalist* 105 (1971) 113–136.
Grayson, D.K. *The Establishment of Human Antiquity* (New York, 1983).
Gregorio, M.A. di. "The Dinosaur Connection: a Reinterpretation of T.H. Huxley's Evolutionary View", *Journal of the History of Biology* 15 (1982) 397–418.
Gregorio, M.A. di. *T.H. Huxley's Place in Natural Science* (New Haven, 1984).
Gruber, J.W. "The Neanderthal Controversy", *Scientific Monthly* 67 (1948) 436–439.
Gruber, J.W. "Brixham Cave and the Antiquity of Man", in: M.E. Spiro, ed. *Context and Meaning in Cultural Anthropology* (New York, 1965) 373–402.
Haeckel, E. *Natürliche Schöpfungsgeschichte: gemeinverständliche wissenschaftliche Vorträge über die Entwickelungslehre im Allgemeinen und Diejenige von Darwin, Goethe und Lamarck im Besonderen* (Berlin, 1868).
Haeckel, E. "Systematische Phylogenie der Wirbelthiere (Vertebrata)", vol. III of: *Systematische Phylogenie: Entwurf eines natürlichen Systems der Organismen auf Grund ihrer Stammesgeschichte* (3 vols; Berlin, 1894–1896).
Haeckel, E. *Unsere gegenwärtige Kenntniss vom Ursprung des Menschen* (Leipzig, 1898).
Haeckel, E. *Das Menschenproblem und die Herrentiere von Linné* (Frankfurt, 1908).
Haller, J.S. *Outcasts from Evolution: Scientific Attitudes of Racial Inferiority, 1859–1900* (Urbana, 1971).
Hammond, M. "The Expulsion of the Neanderthals from Human Ancestry", *Social Studies of Science* 12 (1982) 1–36.
Harvey, J. "Evolutionism Transformed: Positivists and Materialists in the 'Société d'Anthropologie de Paris' from Second Empire to Third Republic", in: D.R. Oldroyd and I. Langham, eds. *The Wider Domain of Evolutionary Thought* (Dordrecht, 1983) 289–310.
Hegeman, J.G. "Nederlandse reacties op de evolutieleer 1860–1875: een terreinverkenning", *Bijdragen en mededelingen betreffende de geschiedenis der Nederlanden* 85 (1970) 261–314.

Hepburn, D. "The Trinil Femur (Pithecanthropus erectus), Contrasted with the Femora of Various Savage and Civilized Races", *Journal of Anatomy and Physiology* 31, N.S. 11 (1897) 1–17.

Herbert, S. "The Place of Man in the Development of Darwin's Theory of Transmutation", *Journal of the History of Biology* 7 (1974) 217–258; 10 (1977) 155–227.

Hofman, M.A. "Encephalization in Mammals in Relation to the Size of the Cerebral Cortex", *Brain, Behavior and Evolution* 20 (1982) 84–96.

Hofman, M.A. "Encephalization in Hominids: Evidence for the Model of Punctuationalism", *Brain, Behavior and Evolution* 22 (1983) 102–117.

Holloway, R.L. "The Evolution of the Primate Brain: Some Aspects of Quantitative Relations", *Brain Research* 7 (1968) 121–172.

Hooijer, D.A. "Fossil Proboscidea from the Malay Archipelago and the Punjab", *Zoölogische verhandelingen (RMNH)* 28 (1955) 1–146.

Houzé, E. "Le Pithecanthropus erectus: discussion", *Revue de l'Université de Bruxelles* 1 (1895–1896) 5–42.

Hrdlička, A. "The Skeletal Remains of Early Man", *Smithsonian Miscellaneous Collections* 83 (1930) 1–379.

Hubrecht, A.A.W. "Aap of mensch?", *De gids* 60 (1896) 507–519.

Huizinga, J. "De prehistorische mens in Nederland", in: J.E.A.T. Bogaers, ed. *Honderd eeuwen Nederland* (Den Haag, 1959) 37–56.

Huxley, J.S. *Problems of Relative Growth* (New York, 1972; reprint of the first edition; London, 1932).

Huxley, J.S. and Teissier, G. "Terminology of Relative Growth", *Nature* 137 (1936) 780–781.

Huxley, T.H. "Man's Place in Nature", in vol. VII of: *Collected Essays, 1893–1894* (9 vols; Hildesheim, 1970).

Idenburg, P.J. *De Leidse universiteit, 1928–1946: vernieuwing en verzet* ('s-Gravenhage, 1978).

Jacob, W. "Virchows Begriff der 'naturwissenschaftlichen Methode': Deutung und Grenzen", in: H. Querner and H. Schipperges, eds. *Wege der Naturforschung, 1822–1972, im Spiegel der Versammlungen Deutscher Naturforscher und Aerzte* (Berlin, 1972) 88–100.

Jaekel, O. Commentary on Dubois' "Pithecanthropus erectus, betrachtet als eine wirkliche Uebergangsform und als Stammform des Menschen", *Zeitschrift für Ethnologie* 27 (1895) 747–748.

Jerison, H.J. *Evolution of the Brain and Intelligence* (New York, 1973).

Kate, H. ten. Review of Dubois' Pithecanthropus erectus, eine menschenähnliche Uebergangsform aus Java, *Nederlandsch koloniaal centraalblad* I (1894–1895) 127–129.

Keibel, F. "Gustav Albert Schwalbe", *Anatomischer Anzeiger* 49 (1916–1917) 210–221.

Keith, A. Commentary on Cunningham's [A Paper on Pithecanthropus erectus], *Journal of Anatomy and Physiology* 29, N.S. 9 (1985) XVIII–XIX.

Keith, A. "Pithecanthropus erectus – a Brief Review of Human Fossil Remains", *Science Progress* 3 (1895) 348–369.

Keith, A. *The Antiquity of Man* (second edition; London, 1925).

Kennedy, K.A.R. "The Phylogenetic Tree: an Analysis of Its Development in Studies of Human Evolution", *Kroeber Anthropological Society Papers* 23 (1960) 7–53.

King, W. "The Neanderthal Skull", *Anthropological Review* 1 (1863) 393–394.

Klaatsch, H. "Der gegenwärtige Stand der Pithecanthropus-Frage", *Zoologisches Zentralblatt* 6 (1899) 217–235.

Klaatsch, H. "Die Stellung des Menschen in der Primatenreihe und der Modus seiner Hervorbildung aus einer niederen Form", *Mittheilungen der Anthropologischen Gesellschaft in Wien* 30 (1900) 88–91.

Klaatsch, H. "Die Aurignac-Rasse und ihre Stellung im Stammbaum der Menschheit", *Zeitschrift für Ethnologie* 42 (1910) 513–577.

Klatt, B. "Zur Methodik vergleichender metrischer Untersuchungen", *Biologisches Zentralblatt* 39 (1919) 406–421.

Koenigswald, G.H.R. von. "Een nieuwe Pithecanthropus ontdekt", *Algemeen Indisch dagblad*, 28 March 1936.

Koenigswald, G.H.R. von. "Erste Mitteilung über einen fossilen Hominiden aus dem Altpleistocän Ostjavas", *Proceedings of the Section of Sciences Koninklijke Nederlandsche Akademie van Wetenschappen* 39 (1936) 1000–1009.

Koenigswald, G.H.R. von. "Ein Unterkieferfragment des Pithecanthropus", *Proceedings of the Section of Sciences Koninklijke Nederlandsche Akademie van Wetenschappen* 40 (1937) 883–893.

Koenigswald, G.H.R. von. "Pithecanthropus received into the Human Family", *Illustrated London News* 191 (1937) 1039–1041.

Koenigswald, G.H.R. von. "Ein neuer Pithecanthropus-Schädel", *Proceedings of the Section of Sciences Koninklijke Nederlandsche Akademie van Wetenschappen* 41 (1938) 185–192.

Koenigswald, G.H.R. von. "Fossil Hominids from the Lower Pleistocene of Java", in: *Proceedings of the International Geological Congress, Great Britain, 1948* (London, 1950) Part IX, 59–61.

Koenigswald, G.H.R. von. *Begegnungen mit dem Vormenschen* (Düsseldorf, 1955).

Koenigswald, G.H.R. von. "De eerste 'aapmensen' in Azie", in: *De evolutie van de mens: de speurtocht naar ontbrekende schakels* (Natuur en techniek; Maastricht, 1981) 238–257.

Koenigswald, G.H.R. von and Weidenreich, F. "The Relationship between Pithecanthropus and Sinanthropus", *Nature* 144 (1939) 926–929.

Kohlbrugge, J.H.F. *Die morphologische Abstammung des Menschen: kritische Studie über die neueren Hypothesen* (Stuttgart, 1908).

Kollmann, J. Commentary on Dubois' "Pithecanthropus erectus, betrachtet als eine wirkliche Uebergangsform und als Stammform des Menschen", *Zeitschrift für Ethnologie* 27 (1895) 740–744.

Koninklijke Natuurkundige Vereeniging. *Een eeuw natuurwetenschap in Indonesië, 1850–1950* (Bandung, 1950).

Kottler, M.J. "Alfred Russel Wallace, the Origin of Man and Spiritualism", *Isis* 65 (1974) 145–192.

Krause, W. Discussion of Dubois' Pithecanthropus erectus, eine menschenähnliche Uebergangsform aus Java, *Zeitschrift für Ethnologie* 27 (1895) 78–81.

Kruytzer, E.M. "Bezoek van het Natuurhistorisch Genootschap aan 'De Bedelaar' op 6 Juni 1938", *Natuurhistorisch maandblad* 27 (1938) 61–65.

Lapicque, L. "Sur la rélation du poids de l'encéphale au poids du corps", *Comptes rendus hebdomadaires des séances de la Société de Biologie* 50 (1898) 63.

Lapicque, L. "Comparaison du poids encéphalique entre les deux sexes de l'espèce humaine", *Comptes rendus des séances de la Société de Biologie* 63 (1907) 434.

Lapicque, L. "Tableau général des poids somatiques et encéphaliques dans les espèces animales", *Bulletins et mémoires de la Société d'Anthropologie de Paris* S. 5, 8 (1907) 248–262.

Lapicque, L. "Le poids encéphalique en function du poids corporel entre individus d'une même espèce", *Bulletins et mémoires de la Société d'Anthropologie de Paris* S. 5, 8 (1907) 313–345.

Lapicque, L. "La grandeur relative de l'oeil et l'appréciation du poids encéphalique", *Comptes rendus hebdomadaires des séances de l'Académie des Sciences* 147 (1908) 209.

Lapicque, L. "Le poids de l'encéphale dans les différents groupes d'oiseaux", *Bulletin du Museum National d'Histoire Naturelle* 15, nr. 7 (1909) 408–412.

Lapicque, L. "Relation du poids encéphalique à la surface rétinienne dans quelques ordres de mammifères", *Comptes rendus hebdomadaires des séances de l'Académie des Sciences* 151 (1910) 1393.

Lapicque, L. "Sur le développement phylogénétique du cerveau (d'après les travaux récents d'Eugène Dubois)", *Annales des sciences naturelles. Zoologie* S. 10, 17 (1934) 369–380.

Lapicque, L. and Dhéré, Ch. "Sur le rapport entre la grandeur du corps et le développement de l'encéphale", *Archives de physiologie normale et pathologique* 30, nr. 4 (1898) 763–773.

Lapicque, L. and Girard, P. "Poids de l'encéphale, en fonction du poids du corps chez les oiseaux", *Comptes rendus hebdomadaires des séances de l'Académie des Sciences* 140 (1905) 1057–1059.

Lartet, E. "Note sur les ossements fossiles des terrains tertiares de Simorre, de Sansan, etc. dans le département du Gers, et sur la découverte de la mâchoire d'un singe fossile", *Comptes rendus hebdomadaires des séances de l'Académie des Sciences* 4 (1837) 85–93.

Lartet, E. "Nouvelles observations sur une mâchoire inférieure fossile, crue d'un singe voisin du gibbon, et sur quelques dents et ossements attribués à d'autres quadrumanes", *Comptes rendus hebdomadaires des séances de l'Académie des Sciences* 4 (1837) 583–584.

Lartet, E. "Note sur un grand singe fossile qui se rattache au groupe des singes supérieurs", *Comptes rendus hebdomadaires des séances de l'Académie des Sciences* 43 (1856) 219–223.

Lemaire, T. "De Bedelaar van prof. Dubois: geschiedenis van een landgoed", *Rondom het Leudal* 2 (1977) nr. 5, 3–6; nr. 6, 19–23; nr. 7, 10–15; nr. 8, 16–21; nr. 9, 10–15; nr. 10, 20–24; nr. 11, 18–22; nr. 12, 20–25.

Lodge, R. "Turner, Sir William", in: H.W.C. Davis and J.R.H. Weaver, eds. *The Dictionary of National Biography, 1912–1921* (Oxford, 1926) 536–539.

Luschan, F. von. Commentary on Krause's discussion of Dubois' Pithecanthropus erectus, eine menschenähnliche Uebergangsform aus Java, *Zeitschrift für Ethnologie* 27 (1895) 81.

Lydekker, R. "Notices of Siwalik Mammals", *Records of the Geological Survey of India* 12 (1879) 33–52.

Lydekker, R. "Siwalik Mammalia", *Memoirs of the Geological Survey of India, Palaeontologia Indica* Series 10, *Indian Tertiary and Post-Tertiary Vertebrata* 4, pt. 1 (1886) 1–18.

L[ydekker], R. Review of Dubois' Pithecanthropus erectus, eine menschenähnliche Uebergangsform aus Java, *Nature* 51 (1895) 291.

Lyell, Ch. *Principles of Geology, Being an Attempt to Explain the Former Changes of the Earth's Surface, by Reference to Causes Now in Operation* (3 vols; London, 1830–1833).

Lyell, Ch. *The Geological Evidences of the Antiquity of Man: with Remarks on Theories of the Origin of Species by Variation* (London, 1863).

Lyon, J. "The Search for Fossil Man: Cinq Personnages à la Recherche du Temps Perdu", *Isis* 61 (1970) 68–84.

Manouvrier, L. "Sur l'interprétation de la quantité dans l'encéphale et dans le cerveau en particulier", *Mémoires de la Société d'Anthropologie de Paris* S. 2, 3 (1885) 137–323.

Manouvrier, L. "Discussion du 'Pithecanthropus erectus', comme précurseur présumé de l'homme", *Bulletins de la Société d'Anthropologie de Paris* 6 (1895) 12–47.

Manouvrier, L. "Le Pithecanthropus", *Revue mensuelle de l'École d'Anthropologie de Paris* 5 (1895) 69–72.

Manouvrier, L. "Deuxième étude sur le Pithecanthropus erectus comme précurseur présumé de l'homme", *Bulletins de la Société d'Anthropologie de Paris* 6 (1895) 553–651.

Manouvrier, L. "Le Pithecanthropus erectus et la théorie transformiste", *Revue scientifique* 5 (1896) 289–299.

Manouvrier, L. "Réponse aux objections contre le Pithecanthropus", *Bulletins de la Société d'Anthropologie de Paris* 7 (1896) 396–460.

Marsh, O.C. "Small Size of the Brain in Tertiary Mammals", *American Journal of Science* 8 (1874) 66–67.

Marsh, O.C. "On the Pithecanthropus erectus, Dubois, from Java", *American Journal of Science* S. 3, 49 (1895) 144–147.

Marsh, O.C. Commentary on Dubois' "Pithecanthropus erectus, eine menschenähnliche Uebergangsform", in: *Compte-rendu des séances du Troisième Congres International de Zoologie, Leyde, 16–21 septembre 1895* (Leiden, 1896) 272.

Marsh, O.C. "On the Pithecanthropus erectus, from the Tertiary of Java", *American Journal of Science* S. 4, 1 (1896) 475–482.

Marsh, O.C. "The Ape-Man from the Tertiary of Java", *Science* N.S. 3 (1896) 789–793.

Martin, K. "Ueberreste vorweltlicher Proboscidier auf Java und Banka", *Sammlungen des Geologischen Reichs-Museums in Leiden* 4 (1884–1889) 1–24.

Martin, K. "Fossile Säugethierreste von Java und Japan", *Sammlungen des Geologischen Reichs-Museums in Leiden* 4 (1884–1889) 25–69.

Martin, K. "Notiz über den angeblich fossilen, menschlichen Unterkiefer vom Caberge bei Maastricht", *Verslagen en mededeelingen der Koninklijke Akademie van Wetenschappen, afdeeling natuurkunde* S. 3, 5 (1889) 434–440.

Martin K. Commentary on Dubois' "Pithecanthropus erectus, eine menschenähnliche Uebergangsform", in: *Compte-Rendu des séances du Troisième Congres International de Zoologie, Leyde, 16–21 septembre 1895* (Leiden, 1896) 273.

Martin, R. "Kritische Bedenken gegen den Pithecanthropus erectus Dubois", *Globus* 67 (1895) 213–217.

Martin, R. *Weitere Bemerkungen zur Pithecanthropus-Frage* (Zürich, 1896).

Matschie, P. "Anthropopithecus erectus E. Dubois", *Naturwissenschaftliche Wochenschrift* 9 (1894) 122–123.

Matschie, P. "Noch einmal Pithecanthropus erectus", *Naturwissenschaftliche Wochenschrift* 10 (1895) 81–82.

Mayr, E. and Provine, W.B., eds. *The Evolutionary Synthesis: Perspectives on the Unification of Biology* (Cambridge, Mass., 1980).

Montgomery, W.M. *Evolution and Darwinism in German Biology, 1800–1883* (Diss. Univ. of Texas at Austin, 1974; Ann Arbor, Mich., 1982).

Mortillet, G. de. "l'Anthropopithèque", *Revue mensuelle de l'École d'Anthropologie de Paris* 2 (1892) 137–154.

Mortillet, G. de. "Précurseur de l'homme et Pithécanthrope", *Revue mensuelle de l'École d'Anthropologie de Paris* 6 (1896) 305–317.

Munro, R. "On Intermediary Links between Man and the Lower Animals", *Proceedings of the Royal Society of Edinburgh* 21 (1897) 249–250.

Nehring, A. "Menschenreste aus einem Sambaqui von Santos in Brasilien, unter Vergleichung der Fossilreste des Pithecanthropus erectus Dubois", *Zeitschrift für Ethnologie* 27 (1895) 710–722.

Nehring, A. Commentary on Dubois' "Pithecanthropus erectus, betrachtet als eine wirkliche Uebergangsform und als Stammform des Menschen", *Zeitschrift für Ethnologie* 27 (1895) 738–740.

Newton, E.T. "Palaeolithic Man", *Nature* 57 (1897–1898) 354–355.

Newton, E.T. "The Evidence for the Existence of Man in the Tertiary Period", *Proceedings of the Geologists Association* 15 (1899) 63–82.

Oakley, K.P. "The Problem of Man's Antiquity: a Historical Survey", *Bulletin of the British Museum (Natural History)* 9 (1963–1964) 85–155.

Odom, H.H. "Generalizations on Race in Nineteenth Century Physical Anthropology", *Isis* 58 (1967) 5–18.

Oldroyd, D.R. *Darwinian Impacts: an Introduction to the Darwinian Revolution* (Milton Keynes, 1980).

Oppenoorth, W.F.F. "Homo (Javanthropus) soloensis, een pleistocene mensch van Java", *Wetenschappelijke mededeelingen Dienst van den Mijnbouw in Nederlandsch-Indië* 20 (1932) 49–63.

Osborn, H.F. *Men of the Old Stone Age: Their Environment, Life and Art* (New York, 1915).

Pekelharing, C.A. "In memoriam Thomas Place", *Nederlandsch tijdschrift voor geneeskunde* 55 (1911) 689–691.

Penniman, T.K. *A Hundred Years of Anthropology* (third edition; London, 1965).

Pettit, A. "Le Pithecanthropus erectus", *l'Anthropologie* 6 (1895) 65–69.

Place, Th. *De descendentieleer* (Amsterdam, 1871).

Polak, B. "In memoriam Eugène Dubois: Eijsden, 28 januari 1858 – Haarlem, waarschijnlijk december 1940", *Natuurwetenschappelijk tijdschrift voor Nederlandsch-Indië* 101 (1941) 89–95.

Poliakov, L. *The Aryan Myth: a History of Racist and Nationalist Ideas in Europe* (London, 1974).

Puydt, M. de and Lohest, M. "l'Homme contemporain du mammouth à Spy", in: *Compte-rendu du Congres de la Fédération Archéologique et Historique de Belgique* (Namur, 1886) 207–235.

Quenstedt, W. and Quenstedt, A. *Hominidae Fossiles* ('s-Gravenhage, 1936; vol 74 of: W. Quenstedt, ed. *Fossilium Catalogus. I: Animalia*).

Rainger, R. *The Understanding of the Fossil Past: Paleontology and Evolution Theory, 1850–1910* (Diss. Indiana Univ., 1982; Ann Arbor, Mich., 1984).

Ranke, J. "Die heutigen und die vorgeschichtlichen Menschenrassen", vol. II of: *Der Mensch* (2 vols; Leipzig, 1894).

Reader, J. *Missing Links; the Hunt for Earliest Man* (London, 1981).

Regteren Altena, C.O. van. "Achttiende-eeuwse verzamelaars van fossielen te Maastricht en het lot hunner collecties", *Publicaties van het Natuurhistorisch Genootschap in Limburg* 9 (1956) 83–112.

Regteren Altena, C.O. van. "Nieuwe gegevens over achttiende-eeuwse verzamelaars van fossielen te Maastricht", *Natuurhistorisch maandblad* 52 (1963) 28–32.

Rietschoten, B.D. van. [Uit een schrijven van den heer van Rietschoten te Blitar], *Natuurkundig tijdschrift voor Nederlandsch-Indië* 48 (1889) 346–347.

Rijnberk, G. van. "Prof. Place", *Amsterdamsche studenten-almanak* 80 (1910) 163–165.

Rosenberg, E. Commentary on Dubois' "Pithecanthropus erectus, eine menschenähnliche Uebergangsform", in: *Compte-rendu des séances du Troisième Congres International de Zoologie, Leyde, 16–21 septembre 1895* (Leiden, 1896) 272.

Rudwick, M.J.S. *The Meaning of Fossils* (second edition; New York, 1976).

Rückert, J. "Johannes Ranke", *Jahrbuch der Königlich Bayerischen Akademie der Wissenschaften* (1917) 67–77.

Russell, E.S. *Form and Function: a Contribution to the History of Animal Morphology* (Chicago 1982; reprint of the first edition; London, 1916; with an introduction by G.V. Lauder).

Schaaffhausen, H. "Ueber Beständigkeit und Umwandlung der Arten", *Verhandlungen des Naturhistorischen Vereins der Preussischen Rheinlande und Westphalens* 10 (1853) 420–451.

Schaaffhausen, H. "Zur Kenntniss der ältesten Rassenschädel", *Archiv für Anatomie, Physiologie und wissenschaftliche Medicin* 25 (1858) 453–478.

Schaaffhausen, H. "On the Crania of the Most Ancient Races of Man. By Prof. D. Schaaffhausen, of Bonn." (From Müller's Archiv., 1858, pp. 453. With remarks and original figures, taken from a cast of the Neanderthal cranium. By George Busk, F.R.S., &c.), *Natural History Review* 1 (1861) 155–176.

Schmidt, H., ed. *Was wir Ernst Haeckel verdanken: ein Buch der Verehrung und Dankbarkeit* (Leipzig, 1914).

Smidt, J. *Die Umweltlehre Jacob von Uexkülls in ihrer Bedeutung für die Entwicklung der vergleichenden Verhaltensforschung* (Marburg, 1980).

Schreuder, A. "Conodontes (Trogontherium) and Castor from the Teglian Clay, Compared with the Castoridae from Other Localities", *Archives du Musée Teyler* S. 3, 6 (1929) 99–321.

Schroevers-Kommandeur, H. and Jager, S. de. "De Henkeput: opgravingen en determinatie van dierlijke resten", *Natuurhistorisch maandblad* 55 (1966) 117–126.

Schuster, J. "Die Flora der Trinil-Schichten", in: M.L. Selenka and M. Blanckenhorn, eds. *Die Pithecanthropus-Schichten auf Java* (Leipzig, 1911).

Schwalbe, G. "Studien über Pithecanthropus erectus Dubois", *Zeitschrift für Morphologie und Anthropologie* 1 (1899) 16–240.

Schwalbe, G. "Ziele und Wege einer vergleichenden physischen Anthropologie", *Zeitschrift für Morphologie und Anthropologie* 1 (1899) 1–15.

Schwalbe, G. "Der Neanderthalschädel", *Bonner Jahrbücher* 106 (1901) 1–72.

Schwalbe, G. *Die Vorgeschichte des Menschen* (Braunschweig, 1904).

Schwalbe, G. and Fischer E., ed. "Studien über das Femur von Pithecanthropus erectus Dubois", *Zeitschrift für Morphologie und Anthropologie* 21 (1921) 289–360.

Schwartz, J.S. "Darwin, Wallace, and the Descent of Man", *Journal of the History of Biology* 17 (1984) 271–289.

Selenka, M.L. [Die fossile Zähne von Trinil], *Tijdschrift van het Koninklijk Nederlands Aardrijkskundig Genootschap te Amsterdam* S. 2, 26 (1909) 398–399.

Selenka, M.L. "Der menschliche Zahn von Trinil: eine Erwiderung", *Korrespondenzblatt der Deutschen Gesellschaft für Anthropologie, Ethnologie und Urgeschichte* 40 (1909) 24.

Selenka, M.L. and Blanckenhorn, M., eds. *Die Pithecanthropus-Schichten auf Java* (Leipzig, 1911).

Sholl, D. "The Quantitative Investigation of the Vertebrate Brain and the Applicability of Allometric Formulae to Its Study", *Proceedings of the Royal Society of London* Series B, 135 (1947–1948) 243–255.

Sirks, M.J. *Indisch natuuronderzoek: een beknopte geschiedenis van de beoefening der natuurwetenschappen in de Nederlandsche koloniën* (Amsterdam, 1915).

Smit, P. "Houding en reactie van het katholieke volksdeel ten aanzien van de evolutiegedachte", *Annalen van het Thijmgenootschap* 68 (1980) 223–240.

Smith, G.E. "The Peking Man: a New Chapter of Human History", *Illustrated London News* 175 (1929) 672–673.

Snell, O. "Das Gewicht des Hirnes und des Hirnmantels der Säugethiere in Beziehung zu deren geistigen Fähigkeiten", *Sitzungsberichte der Gesellschaft für Morphologie und Physiologie in München* 7 (1892) 90–94.

Snell, O. "Die Abhängigkeit des Hirngewichtes von dem Körpergewicht und den geistigen Fähigkeiten", *Archiv für Psychiatrie und Nervenkrankheiten* 23 (1892) 436–446.

Sollas, W.J. "'Pithecanthropus erectus' and the Evolution of the Human Race", *Nature* 53 (1895–1896) 150–151.

"Special Section on American Morphology at the Turn of the Century", *Journal of the History of Biology* 14 (1981) 83–191.

Starck, D. "Die idealistische Morphologie und ihre Nachwirkungen", *Medizinhistorisches Journal* 15 (1980) 44–56.

Stebbins, R.E. *French Reactions to Darwin, 1859–1882* (Diss. University of Minnesota, 1965; Ann Arbor, Mich., 1983).

Stewart, T.D. "The Effect of Darwin's Theory of Evolution on Physical Anthropology", in: B.J. Meggers, ed. *Evolution and Anthropology: a Centennial Appraisal* (Washington, 1959) 11–25.

Stocking, G. "The Persistence of Polygenist Thought in Post-Darwinian Anthropology", in: G. Stocking, ed. *Race, Culture and Evolution: Essays in the History of Anthropology* (New York, 1968) 42–68.

Suringar, W.F.R. *Zakflora: handleiding tot het bepalen van de in Nederland wildgroeiende planten* (Leeuwarden, 1870).

Tesch, P and Brongersma, L.D. "Eugène Dubois", *Geologie en mijnbouw* 3 (1941) 29–33.

Thompson, d'Arcy W. *On Growth and Form* (Cambridge, 1917).

Tillema, J.A.C. *Victor de Stuers: ideeën van een individualist* (Assen, 1982).

Tobias, P.V. "The Life and Work of Professor Dr. G.H.R. von Koenigswald", in: *Auf den Spuren des Pithecanthropus: Leben und Werk von Professor Dr. Gustav Heinrich Ralph von Koenigswald (1902–1982). Vorträge einer akademischen Gedächtnisfeier am 20. November 1983 im Senckenberg-Museum, Frankfurt am Main* (Frankfurt am Main, 1984) 25–96.

Todes, D.P. "V.O. Kovalevskii: the Genesis, Content, and Reception of His Paleontological Work", *Studies in the History of Biology* 2 (1978) 99–165.

Topinard, P. Review of Turner's "On M. Dubois' Description of Remains Recently Found in Java, Named by Him Pithecanthropus erectus. With Remarks on So-Called Transitional Forms between Apes and Man", *l'Anthropologie* 6 (1895) 605–607.

Turner, W. "On M. Dubois' Description of Remains Recently Found in Java, Named by Him Pithecanthropus erectus. With Remarks on So-Called Transitional Forms between Apes and Man", *Journal of Anatomy and Physiology* 29, N.S. 9 (1895) 424–445.

Ubaghs, M.C. *Les ateliers ou stations, dits préhistoriques de St.-Gertrude et Rijckholt, près de Maestricht* (Liège, 1887).

Uexküll, J. von. *Theoretische Biologie* (Berlin, 1920).

Uexküll, J. von. *Umwelt und Innenwelt der Tiere* (second edition; Berlin, 1921).

Uschmann, G. *Geschichte der Zoologie und der zoologischen Anstalten in Jena, 1779–1919* (Jena, 1959).

Valk, J.P. de. "De katholieken en het onderwijs", *Spiegel historiael* 13 (1978) 736–743.

Veer, P.H.W.A.M. de. *Leven en werk van Hugo de Vries* (Groningen, 1969).

Verneau, R. "Encore le Pithecanthropus erectus", *l'Anthropologie* 6 (1895) 725–726.

Versluys, J. *Hirngrösse und hormonales Geschehen bei der Menschwerdung: ein*

Vortrag über die Untersuchungen von E. Dubois und L. Bolk und die Bedeutung derselben (Wien 1939; with contributions by O. Pötzl and K. Lorenz).

Virchow, R. Commentary on Krause's discussion of Dubois' Pithecanthropus erectus, eine menschenähnliche Uebergangsform aus Java, *Zeitschrift für Ethnologie* 27 (1895) 81–87, 88.

Virchow, R. [Die Frage von dem Pithecanthropus erectus], *Zeitschrift für Ethnologie* 27 (1895) 435–442.

Virchow, R. [Weitere Mittheilungen über den Pithecanthropus erectus Dub.], *Zeitschrift für Ethnologie* 27 (1895) 648–657.

Virchow, R. "Der Pithecanthropus vor dem zoologischen Kongress zu Leiden", *Die Nation: Wochenschrift für Politik, Volkswirtschaft und Litteratur* 12, nr. 4 (1895) 53–55.

Virchow, R. Commentary on Dubois' "Pithecanthropus erectus, betrachtet als eine wirkliche Uebergangsform und als Stammform des Menschen", *Zeitschrift für Ethnologie* 27 (1895) 744–747, 748–749.

Virchow, R. "Exostosen und Hyperostosen von Extremitätenknochen des Menschen, im Hinblick auf den Pithecanthropus", *Zeitschrift für Ethnologie* 27 (1895) 787–793.

Visser, R.P.W. "De Utrechtse zoölogie rond 1900: van morfologie naar vergelijkende fysiologie", *Tijdschrift voor de geschiedenis der geneeskunde, natuurwetenschappen, wiskunde en techniek* 7 (1984) 49–59.

Vogt, C. *Lectures on Man, his Place in Creation, and in the History of the Earth* (London, 1864).

Volz, W. "Pithecanthropus erectus Dub.: eine menschenähnliche Uebergangsform aus Java", *Jahresbericht der Schlesischen Gesellschaft für Vaterländische Cultur* 74 (1896) 5–8.

Volz, W. "Demonstration eines Schädel-Abgusses von Pithecanthropus erectus", *Jahresbericht der Schlesischen Gesellschaft für Vaterländische Cultur* 75 (1897) 10–20.

Volz, W. "Das geologische Alter der Pithecanthropus-Schichten bei Trinil, Ost-Java", *Neues Jahrbuch für Mineralogie, Geologie und Palaeontologie* (1907) 256–271.

Vos, J. de and Sondaar, P.Y. "The Importance of the 'Dubois Collection' Reconsidered", *Modern Quaternary Research in Southeast Asia* 7 (1982) 35–63.

Vries, J.J. de. *Anderhalve eeuw hydrologisch onderzoek in Nederland: overzicht van de ontwikkeling van de wetenschappelijke kennis van het water in Nederland tussen 1830–1980, beschouwd vanuit een geohydrologische gezichtshoek* (Amsterdam, 1982).

Waldeyer, W. Commentary on Krause's discussion of Dubois' Pithecanthropus erectus, eine menschenähnliche Uebergangsform aus Java, *Zeitschrift für Ethnologie* 27 (1895) 88.

Walkhoff, O. "Der Zahn von Sondé (ein fossiler Menschenzahn von Java)", in: M.L. Selenka and M. Blanckenhorn, eds. *Die Pithecanthropus-Schichten auf Java* (Leipzig, 1911) 214–221.

Wallace, A.R. "On the Origin of Human Races and the Antiquity of Man Deduced from the Theory of 'Natural Selection'", *Journal of the Anthropological Society of London* 2 (1864) 158–183.

Wallace, A.R. *Contributions to the Theory of Natural Selection with Some of Its Applications* (London, 1870).

Walther, J. and Heberer, G., eds. *Im Banne Ernst Haeckels: Jena um die Jahrhundertwende* (Göttingen, 1953).

Weber, M. *Studien über Säugethiere: ein Beitrag zur Frage nach dem Ursprung der Cetaceen* (Jena, 1886).

Weber, M. "Vorstudien über das Hirngewicht der Säugethiere", in vol. III of: *Festschrift zum siebenzigsten Geburtstage von C. Gegenbaur am 21. August 1896* (3 vols; Leipzig, 1896–1897) 105–123.

Weber, M. "Over het hersengewicht der zoogdieren", *Verslagen van de gewone vergaderingen der wis- en natuurkundige afdeeling der Koninklijke Akademie van Wetenschappen te Amsterdam* 5 (1897) 153–156.

Weidenreich, F. "Sinanthropus pekinensis und seine Bedeutung für die Abstammungsgeschichte des Menschen", *Die Naturwissenschaften* 19 (1931) 817–825.

Weidenreich, F. "Sinanthropus pekinensis and Its Position in the Line of Human Evolution", *Peking Natural History Bulletin* 10 (1936) 281–290.

Weidenreich, F. "The New Discoveries of Sinanthropus pekinensis and Their Bearing on the Sinanthropus and Pithecanthropus Problems", *Bulletin of the Geological Society of China* 16 (1936–1937) 439–470.

Weidenreich, F. "The New Discovery of Three Skulls of Sinanthropus pekinensis", *Nature* 139 (1937) 269–272.

Weidenreich, F. "Pithecanthropus and Sinanthropus", *Nature* 141 (1938) 378–379.

Weidenreich, F. "Discovery of the Femur and the Humerus of Sinanthropus pekinensis", *Nature* 141 (1938) 614–617.

Weidenreich, F. "The Skull of Sinanthropus pekinensis: a Comparative Study of a Primitive Hominid Skull", *Palaeontologia Sinica* New Series D, nr. 10 (1943) 1–184.

Weinert, H. "Der 'Sinanthropus pekinensis' als Bestätigung des Pithecanthropus erectus", *Zeitschrift für Morphologie und Anthropologie* 29 (1931) 159–187.

Weinert, H. *Menschen der Vorzeit: ein Ueberblick über die altsteinzeitlichen Menschenreste* (second edition; Stuttgart, 1947).

Wendt, H. *Ich suchte Adam* (Hamm, 1953).

White, J.F. and Gould, S.J. "Interpretation of the Coefficient in Allometric Equation", *American Naturalist* 99 (1965) 5–18.

Wirz, K. "Studien über die Cerebralisation: zur quantitativen Bestimmung der Rangordnung bei Säugetieren", *Acta Anatomica* 9 (1950) 134–196.
Zacharias, K.L. *The Construction of a Primate Order: Taxonomy and Comparative Anatomy in Establishing the Human Place in Nature, 1735–1916* (Diss. Johns Hopkins Univ., 1980; Ann Arbor, Mich., 1982).

List of illustrations

INDEX

GPSR Compliance

The European Union's (EU) General Product Safety Regulation (GPSR) is a set of rules that requires consumer products to be safe and our obligations to ensure this.

If you have any concerns about our products, you can contact us on ProductSafety@springernature.com

In case Publisher is established outside the EU, the EU authorized representative is:

Springer Nature Customer Service Center GmbH
Europaplatz 3
69115 Heidelberg, Germany

Batch number: 10823775

Printed by Printforce, the Netherlands